ELEMENTARY

ALGEBRA:

EMBRACING

THE FIRST PRINCIPLES

OF

THE SCIENCE.

BY CHARLES DAVIES, LL.D

AUTHOR OF

ARITHMETIC, ELEMENTARY GEOMETRY, ELEMENTS OF SURVEYING, ELEMENTS OF DESCRIPTIVE AND ANALYTICAL GEOMETRY, ELEMENTS OF DIFFERENTIAL AND INTEGRAL CALCULUS, AND A TREATISE ON SHADES, SHADOWS, AND PERSPECTIVE.

NEW YORK:
PUBLISHED BY A. S. BARNES & CO.
CINCINNATI:—H. W. DERBY & CO
1850.

PREFACE.

ALTHOUGH Algebra naturally follows Arithmetic in a course of scientific studies, yet the change from numbers to a system of reasoning entirely conducted by letters and signs is rather abrupt and not unfrequently discourages the pupil.

In this work it has been the intention to form a connecting link between Arithmetic and Algebra, to unite and blend, as far as possible, the reasoning on numbers with the more abstruse method of analysis.

The Algebra of M. Bourdon has been closely followed. Indeed, it has been a part of the plan, to furnish an introduction to that admirable treatise, which is justly considered, both in Europe and this country, as the best work on the subject of which it treats, that has yet appeared.

This work, however, even in its abridged form, is too voluminous for schools, and the reasoning is too elaborate and metaphysical for beginners.

It has been thought that a work which should so far modify the system of M. Bourdon as to bring it within the scope of our common schools, by giving to it a more practical and tangible form, could not fail to be useful. Such is the object of the ELEMENTARY ALGEBRA.

Having within the past year carefully revised the Algebra of M. Bourdon, and made therein many important changes and alterations, both by the addition of new rules and in the abridgment and simplification of those before given, it became necessary to make corresponding changes in the introductory work. The alterations before made, in the form of an Introduction, the Treatise on Logarithms, and the Supplement containing practical examples, with solutions given in the Key, have all been retained; and the work is now presented to the public in a form which it is hoped will not require alteration.

WEST POINT, *January*, 1845.

DAVIES'
COURSE OF MATHEMATICS.

DAVIES' TABLE BOOK.

DAVIES' FIRST LESSONS IN ARITHMETIC—

DAVIES' ARITHMETIC.—Designed for the use of Academies and Schools.

KEY TO DAVIES' ARITHMETIC.

DAVIES' UNIVERSITY ARITHMETIC—Embracing the Science of Numbers, and their numerous applications.

KEY TO DAVIES' UNIVERSITY ARITHMETIC.

DAVIES' ELEMENTARY ALGEBRA—Being an Introduction to the Science, and forming a connecting link between ARITHMETIC and ALGEBRA.

KEY TO DAVIES' ELEMENTARY ALGEBRA.

DAVIES' ELEMENTARY GEOMETRY.—This work embraces the elementary principles of Geometry. The reasoning is plain and concise, but at the same time strictly rigorous.

DAVIES' ELEMENTS OF DRAWING AND MENSURATION —Applied to the Mechanic Arts.

DAVIES' BOURDON'S ALGEBRA—Including Sturms' Theorem,— Being an Abridgment of the work of M. Bourdon, with the addition of practical examples

DAVIES' LEGENDRE'S GEOMETRY AND TRIGONOMETRY —Being an Abridgment of the work of M. Legendre, with the addition of a Treatise on MENSURATION OF PLANES AND SOLIDS, and a Table of LOGARITHMS and LOGARITHMIC SINES.

DAVIES' SURVEYING—With a description and plates of the THEODOLITE, COMPASS, PLANE-TABLE, and LEVEL: also, Maps of the TOPOGRAPHICAL SIGNS adopted by the Engineer Department—an explanation of the method of surveying the Public Lands, and an Elementary Treatise on NAVIGATION.

DAVIES' ANALYTICAL GEOMETRY—Embracing the EQUATIONS OF THE POINT AND STRAIGHT LINE—of the CONIC SECTIONS—of the LINE AND PLANE IN SPACE—also, the discussion of the GENERAL EQUATION of the second degree, and of SURFACES of the second order.

DAVIES' DESCRIPTIVE GEOMETRY,—With its application to SPHERICAL PROJECTIONS.

DAVIES' SHADOWS AND LINEAR PERSPECTIVE.

DAVIES' DIFFERENTIAL AND INTEGRAL CALCULUS.

CONTENTS.

CHAPTER I.

PRELIMINARY DEFINITIONS AND REMARKS.

CHAPTER II.

ALGEBRAIC FRACTIONS.

CHAPTER III

EQUATIONS OF THE FIRST DEGREE.

CHAPTER IV.

OF POWERS.

CHAPTER V.

CHAPTER VI.

CHAPTER VII.

CHAPTER VIII.

INTRODUCTION.

LESSON I.

1. John and Charles have twelve apples between them, and each has as many as the other: How many has each?

If we suppose the apples divided into two equal parts, it is plain that John will have one part and Charles the other: hence, they will each have six apples.

In Algebra, we often represent numbers by the letters of the alphabet; that is, we take a letter to stand for a number. Thus, let x stand for the apples which John has. Then, as Charles has an equal number, x will also stand for the apples which he has. But together, they have twelve apples; hence, twice x must be equal to 12. This we write thus:

$$x+x=2x=12;$$

and if twice x is equal to 12, it follows that once x, or x, will be equal to 6. This we write thus:

$$x=\frac{12}{2}=6.$$

When we write x by itself, we mean one x, or the same as $1x$. If we write $2x$, we mean that x is twice taken; if $3x$, that it is taken three times, &c.

QUEST.—**1.** In the first question, how many apples has each boy? By what are numbers represented in Algebra? If x stands by itself, how many times x are expressed? What does $2x$ denote? What $3x$? What $4x$, &c. If we have $x+x$, to how many times x is it equal? If we have the value of $2x$, how do we find the value of x?

2. James and John together have 24 peaches, and one has as many as the other: How many has each?

Let x stand for the number of peaches which James has: then x will also be equal to the number of peaches which John has; and since they have 24 between them,

$$x+x=24;$$

that is, $2x=24$ and $x=\frac{24}{2}=12.$

Therefore, each has twelve peaches.

3. William and John have 36 pears, and one has as many as the other: How many has each?

Let the number which each has be denoted by x.

then $x+x=36;$

that is, $2x=36$ and $x=\frac{36}{2}=18.$

4. What number is that which added to itself will give a sum equal to 20?

Let the number be denoted by x: then, as the number is to be added to itself, we have

$$x+x=20;$$

that is, $2x=20$ or $x=\frac{20}{2}=10.$

Hence, 10 is the number.

5. What number is that which added to itself will give a sum equal to 30?

QUEST.—**2.** In the second question, what does x stand for? What is twice x equal to? How then do you find the value of x? **3.** In the third question, what does x stand for? What is x equal to? How do you find the value of x? **4.** In the fourth question, what does x stand for? What is twice x equal to? How do you then find x? **5.** In the fifth question, what does x stand for? How do you find its value?

6 What number is that which added to itself will give a sum equal to 50?

7. What number is that which added to itself will give a sum equal to 100?

8. What number added to itself will give a sum equal to 80.

9. What number added to itself will give a sum equal to 25.

10. What number added to itself will give a sum equal to $37\frac{1}{2}$.

LESSON II.

1. JOHN and Charles together have 12 apples, and Charles has twice as many as John: How many has each?

If we now suppose the apples to be divided into three equal parts, it is evident that John will have one of the parts and Charles two.

Let us denote by x the apples which John has. Then, $2x$ will be equal to what Charles has, and $x+2x$ will be equal to all the apples. This equality is thus expressed:

$$x+2x=12;$$

that is, $$3x=12 \quad \text{or} \quad x=\frac{12}{3}=4$$

therefore, John has 4 apples, and Charles 8.

QUEST.—**6**. How do you find the value of x in the 6th question? **8**. How in the 8th? **9**. How in the 9th? **10**. How in the 10th?

QUESTIONS ON LESSON II.—**1**. Into how many parts may we suppose the 12 apples to be divided? How many of the parts will John have? What is the value of each part? If x stands for one of the parts, what will stand for two parts? What for three parts? If you have the value of $3x$, how will you find the value of x?

2. James and John have 30 pears, and John has twice as many as James: How many has each?

Here, again, let us suppose the whole number to be divided into three equal parts, of which James must have one part, and John two.

Let us then denote by x, the number of pears which James has: then $2x$ will be equal to the number of pears which John has, and $x+2x$ will be equal to the whole number of pears: and we shall have

$$x+2x=30;$$

that is, $$3x=30 \quad \text{or} \quad x=\frac{30}{3}=10.$$

3. William and John have 48 quills between them, and John has twice as many as William: How many has each?

Let the number of quills which William has be denoted by x: then, since John has twice as many, his will be denoted by $2x$, and the quills possessed by both of them, by $x+2x$: Hence, we shall have

$$x+2x=48;$$

that is, $$3x=48 \quad \text{or} \quad x=\frac{48}{3}=16.$$

Hence, William has 16 quills, and John 32.

4. What number is that which added to twice itself, will give a sum equal to 60?

Let the number sought be denoted by x, then twice the number will be denoted by $2x$, and we shall have

$$x+2x=60;$$

that is, $$3x=60 \quad \text{or} \quad x=\frac{60}{3}=20;$$

and we see that 20 added to twice itself will give 60.

QUEST.—**2.** In question second, what is the value of one of the parts? **3.** What in question 3rd? **4.** How do you state question 4th?

5. John says to Charles, "give me your marbles and I shall have three times as many as I have now." "No," says Charles, "but give me yours, and I shall have just 51." How many had each?

Let the number of marbles which John has be denoted by x: then, $2x$ will denote the number which Charles has, and since they have 51 in all, we write

$$x+2x=51;$$

that is, $3x=51$ or $x=\frac{51}{3}=17.$

6. What number is that which added to twice itself will give a sum equal to 75?

Let the number be denoted by x: then, twice the number will be expressed by $2x$, and

$$x+2x=75;$$

that is, $3x=75$ and $x=\frac{75}{3}=25.$

7. What number added to twice itself will give a sum equal to 90?

8. What number added to twice itself will give a sum equal to 57?

9. What number added to twice itself will give a sum equal to 39?

10. What number added to twice itself will give a sum equal to 21?

LESSON III.

1. If James and John together have 24 quills, and John has three times as many as James, how many will each have?

QUEST.—**5.** How do you state question 5th? **6.** Explain the 6th question? **7.** Also the 7th. **8.** What is the required number in the 8th? **9.** What in the 9th? **10.** What in the 10th?

It is plain that if we suppose the twenty-four quills to be divided in four equal parts, that James will have one of the parts, and John three.

Let us now designate by x the number of quills which James has: then $3x$ will denote the number of quills which John has, and we shall have

$$x+3x=24;$$

that is, $4x=24$ and $x=\frac{24}{4}=6$.

2. What number is that which added to three times itself will give a sum equal to 48?

If we denote the number by x, we shall have

$$x+3x=48;$$

that is, $4x=48$ and $x=\frac{48}{4}=12$.

3. John and Charles have 60 apples between them, and Charles has three times as many as John: How many has each?

If we suppose the number of apples to be divided into four equal parts, it is evident that John will have one of those parts, and Charles three.

Let $x=$ the apples which John has: then $3x$ will stand for the apples which Charles has, and we shall have

$$x+3x=60;$$

that is, $4x=60$ and $x=\frac{60}{4}=15$.

Hence, John will have 15 and Charles 45.

QUEST.—**1.** If the 24 quills be divided into four equal parts, how many parts will John have? How many will James have? What is each part equal to? **2.** If three times a number be added to the number, how many times will the number be taken? If $4x$ is equal to 48, what is the value of x? **3.** Explain the third question? If $4x$ is equal to 60, how do you find the value of x?

4. What number is that which being added to three times itself will give a sum equal to 100?

Let the number be denoted by x: then

$$x+3x=100;$$

that is, $4x=100$ and $x=\frac{100}{4}=25$.

5. What number is that which if added to four times itself, the sum will be equal to 60?

Let x denote the number. Then,

$$x+4x=60;$$

that is, $5x=60$ and $x=\frac{60}{5}=12$.

6. What number is that which being multiplied by 3, and the product added to twice the number will give a sum equal to 75?

Let the number be denoted by x.

Then, $3x=$ the product of the number by 3;

and $2x=$ twice the number;

and $3x+2x=5x=75$;

and $x=\frac{75}{5}=15$, the required number.

7. What number is that which being added to three times itself will give a sum equal to 140?

8. What number is that which being multiplied by 5, and the product added to the number, will give a sum equal to 240?

9. What number is that which being multiplied by 2, and then by 3, and the products added together, will give a sum equal to 125?

QUEST.—**5.** If a number be added to four times itself, how many times will the number be taken? **6.** If x stands for any number, what will stand for three times that number? What for twice the number? **7.** Explain the 7th question? How do you state it? What is $4x$ equal to? Why? How then do you find x? **8.** How do you state the 8th question? What is $6x$ equal to? How then do you find x? **9.** If x denotes a number, what will stand for twice the number? What for three times the number?

LESSON IV.

1. John and Charles together have 80 apples, and Charles has four times as many as John: How many has each?

If we suppose the 80 apples to be divided into 5 equal parts, it is evident that John will have one of the parts, and Charles four.

Let x stand for the apples which John has: then $4x$ will stand for the apples which Charles has; and

$$x+4x=80;$$

that is, $5x=80$ and $x=\frac{80}{5}=16.$

2. What number added to four times itself will give a sum equal to 90?

3. What number added to five times itself will give a sum equal to 120?

4. What number added to six times itself will give a sum equal to 245?

5. What number added to seven times itself will give a sum equal to 360?

N. B. The questions in the preceding Lessons will give some idea of the questions to which Algebra may be applied.

QUEST.—**1.** If x stands for John's apples, what will denote Charles'? What will stand for the apples which they both have? If $5x$ is equal to 80, what will x be equal to? **2.** If a number be added to 4 times itself, how many times will the number be taken? If 5 times a number is equal to 90, what is the value of the number? **3.** Explain example 3rd. **4.** Explain question 4th? What does x stand for? **5.** Explain the 5th question?

ELEMENTARY

ALGEBRA.

CHAPTER I.

Preliminary Definitions and Remarks.

1. QUANTITY is a general term applied to every thing which can be estimated or measured.

2. MATHEMATICS is the science of quantity.

3. ALGEBRA is that branch of mathematics in which the quantities considered are represented by letters, and the operations to be performed upon them are indicated by signs. These letters and signs are called symbols.

4. The sign $+$, is called *plus;* and indicates the addition of two or more quantities. Thus, $9+5$, is read, 9 *plus* 5, or 9 augmented by 5.

If we represent the number nine, by the letter a, and the number 5 by the letter b, we shall have $a+b$, which is read, a plus b; and denotes that the number represented by a is to be added to the number represented by b.

5. The sign $-$, is called *minus;* and indicates that one

QUEST.—**1**. What is quantity? **2**. What is Mathematics? **3**. What is Algebra? What are these letters and signs called? **4**. What does the sign plus indicate? **5**. What does the sign minus indicate?

quantity is to be subtracted from another. Thus, $9-5$ is read, 9 *minus* 5, or 9 diminished by 5.

In like manner, $a-b$, is read, a minus b, or a diminished by b.

6. The sign $\times$, is called the sign of *multiplication;* and when placed between two quantities, it denotes that they are to be multiplied together. The multiplication of two quantities is also frequently indicated by simply placing a point between them. Thus, 36×25, or 36.25, is read, 36 multiplied by 25, or the product 36 by 25.

7. The multiplication of quantities, which are represented by letters, is indicated by simply writing them one after the other, without interposing any sign.

Thus, ab signifies the same thing as $a\times b$, or as $a.b$; and abc the same as $a\times b\times c$, or as $a.b.c.$ Thus, if we suppose $a=36$, and $b=25$, we have

$$ab=36\times25=900.$$

Again, if we suppose $a=2$, $b=3$ and $c=4$, we have

$$abc=2\times3\times4=24.$$

It is most convenient to arrange the letters of a product in alphabetical order.

8. In the product of several letters, as abc, the single letters, a, b, and c, are called *factors* of the product. Thus, in the product ab, there are two factors, a and b; in the product abc, there are three, a, b, and c.

QUEST.—6. What is the sign of multiplication? What does the sign of multiplication indicate? In how many ways may multiplication be expressed? 7. If letters only are used, how may their multiplication be expressed? 8. In the product of several letters, what is each letter called? How many factors in ab?—In abc?—In $abcd$?—In $abcdf$?

9. There are three signs used to denote *division*. Thus,

$a \div b$ denotes that a is to be divided by b.

$\frac{a}{b}$ denotes that a is to be divided by b.

$a|\underline{b}$ denotes that a is to be divided by b.

10. The sign $=$, is called the sign of *equality*, and is read, *is equal to*. When placed between two quantities, it denotes that they are equal to each other. Thus, $9-5=4$: that is, 9 minus 5 is equal to 4: Also, $a+b=c$, denotes that the sum of the quantities a and b is equal to c.

If we suppose $a=10$, and $b=5$, we have

$$a+b=c, \quad \text{and} \quad 10+5=c=15.$$

11. The sign $>$, is called the sign of *inequality*, and is used to express that one quantity is greater or less than another.

Thus, $a>b$ is read, a greater than b; and $a<b$ is read, a less than b; that is, the opening of the sign is turned towards the greater quantity. Thus, if $a=9$, and $b=4$, we write, $9>4$.

12. If a quantity is added to itself several times as $a+a+a+a+a+a$, we generally write it but once, and then place a number before it to show how many times it is taken. Thus,

$$a+a+a+a+a=5a.$$

QUEST.—**9.** How many signs are used in division? What are they? **10.** What is the sign equality? When placed between two quantities, what does it indicate? **11.** For what is the sign of inequality used? Which quantity is placed on the side of the opening? **12.** What is a co-efficient? How many times is ab taken in the expression ab? In $3ab$? In $4ab$? In $5ab$? In $6ab$? If no co-efficient is written, what co-efficient is understood?

The number 5 is called the *co-efficient* of a, and denotes that a is taken 5 times.

When the co-efficient is 1 it is generally omitted. Thus, a and $1a$ are the same, each being equal to a, or to one a.

13. If a quantity be multiplied continually by itself, as $a \times a \times a \times a \times a$, we generally express the product by writing the letter once, and placing a number to the right of, and a little above it: thus,

$$a \times a \times a \times a \times a = a^5.$$

The number 5 is called the *exponent* of a, and denotes the number of times which a enters into the product as a factor. For example, if we have a^3, and suppose $a=3$, we write,

$$a^3 = a \times a \times a = 3^3 = 3 \times 3 \times 3 = 27.$$

If $a=4$, $\quad a^3 = 4^3 = 4 \times 4 \times 4 = 64$,

and for $a=5$, $\quad a^3 = 5^3 = 5 \times 5 \times 5 = 125$.

If the exponent is 1 it is generally omitted. Thus, a^1 is the same as a, each expressing a to the first power.

14. The *power* of a quantity is the product which results from multiplying the quantity by itself. Thus, in the example

$$a^3 = 4^3 = 4 \times 4 \times 4 = 64,$$

64 is the third power of 4, and the exponent 3 shows the *degree* of the power.

15. The sign $\sqrt{}$, is called the radical sign, and when

QUEST.—**13.** What does the exponent of a letter show? How many times is a a factor in a^2? In a^3? In a^4? In a^6? If no exponent is written, what exponent is understood? **14.** What is the power of a quantity? What is the third power of 2? Express the 4th power of a. **15.** Express the square root of a quantity. Also the cube root. Also the 4th root.

prefixed to a quantity, indicates that its root is to be extracted Thus,

$\sqrt[2]{a}$ or simply $\sqrt{a}$ denotes the square root of a.

$\sqrt[3]{a}$ denotes the cube root of a.

$\sqrt[4]{a}$ denotes the fourth root of a.

The number placed over the radical sign is called the *index* of the root. Thus, 2 is the index of the square root, 3 of the cube root, 4 of the fourth root, &c.

If we suppose $a=64$, we have

$$\sqrt{64}=8, \quad \sqrt[3]{64}=4.$$

16. Every quantity written in algebraic language, that is, with the aid of letters and signs, is called an *algebraic quantity*, or the *algebraic expression* of a quantity. Thus,

$3a$	is the algebraic expression of three times the number a;
$5a^2$	is the algebraic expression of five times the square of a;
$7a^3b^2$	is the algebraic expression of seven times the product of the cube of a by the square of b;
$3a-5b$	is the algebraic expression of the difference between three times a and five times b;
$2a^2-3ab+4b^2$	is the algebraic expression of twice the square of a, diminished by three times the product of a by b, augmented by four times the square of b.

1. Write three times the square of a multiplied by the cube of b. *Ans.* $3a^2b^3$.

QUEST.—**16**. What is an algebraic quantity? Is $5ab$ an algebraic quantity? Is $9a$? Is $4y$? Is $3b-x$?

2. Write nine times the cube of a multiplied by b, diminished by the square of c multiplied by d. *Ans.* $9a^3b-c^2d$.

3. If $a=2$, $b=3$, and $c=5$, what will be the value of $3a^2$ multiplied by b^2 diminished by a multiplied by b multiplied by c. We have

$$3a^2b^2-abc=3\times2^2\times3^2-2\times3\times5=78.$$

4. If $a=4$, $b=6$, $c=7$, $d=8$, what is the value of $9a^2+bc-ad$? *Ans.* 154.

5. If $a=7$, $b=3$, $c=7$, $d=1$, what is the value of $6ad+3b^2c-4d^2$? *Ans.* 227

6. If $a=5$, $b=6$, $c=6$, $d=5$, what is the value of $9abc-8ad+4bc$? *Ans.* 1564.

7. Write ten times the square of a into the cube of b into c square into d^3.

17. When an algebraic quantity is not connected with any other, by the sign of addition or subtraction, it is called a *monomial*, or a quantity composed of a single term, or simply, a *term*. Thus,

$$3a, \quad 5a^2, \quad 7a^3b^2,$$

are monomials, or single terms.

18. An algebraic expression composed of two or more parts, separated by the sign + or —, is called a *polynomial*, or quantity involving two or more terms. For example,

$$3a-5b \quad \text{and} \quad 2a^2-3cb+4b^2$$

are polynomials.

19. A polynomial composed of two terms, is called a *binomial;* and a polynomial of three terms is called a *trinomial*.

QUEST.—**17.** What is a monomial? Is $3ab$ a monomial? **18.** What is a polynomial? Is $3a-b$ a polynomial? **19.** What is a binomial? What is a trinomial?

20. Each of the literal factors which compose a term is called a *dimension* of this term: and the *degree* of a term is the number of these factors or dimensions. Thus,

$3a$ is a term of one dimension, or of the first degree.

$5ab$ is a term of two dimensions, or of the second degree.

$7a^3bc^2=7aaabcc$ is of six dimensions, or of the sixth degree.

21. A polynomial is said to be *homogeneous*, when all its terms are of the same degree. The polynomial

$3a-2b+c$ is of the first degree and homogeneous.

$-4ab+b^2$ is of the second degree and homogeneous.

$5a^2c-4c^3+2c^2d$ is of the third degree and homogeneous.

$8a^3+4ab+c$ is not homogeneous.

22. A vinculum or bar ———, or a parenthesis ( ), is used to express that all the terms of a polynomial are to be considered together. Thus,

$$\overline{a+b+c}\times b, \text{ or } (a+b+c)\times b,$$

denotes that the trinomial $a+b+c$ is to be multiplied by b;

also, $\overline{a+b+c}\times\overline{c+d+f}$, or $(a+b+c)\times(c+d+f)$,

denotes that the trinomial $a+b+c$ is to be multiplied by the trinomial $c+d+f$.

When the parenthesis is used, the sign of multiplication is usually omitted. Thus,

$$(a+b+c)\times b \text{ is the same as } (a+b+c)b.$$

QUEST.—**20.** What is the dimension of a term? What is the degree of a term? How many factors in $3abc$? Which are they? What is its degree? **21.** When is a polynomial homogeneous? Is the polynomial $2a^3b+3a^2b^2$ homogeneous? Is $2a^4b-b^3$? **22.** For what is the vinculum or bar used? Can you express the same with the parenthesis?

23. The terms of a polynomial which are composed of the same letters, the same letters in each being affected with like exponents, are called *similar terms.*

Thus, in the polynomial

$$7ab+3ab-4a^3b^2+5a^3b^2,$$

the terms $7ab$, and $3ab$, are similar: and so also are the terms $-4a^3b^2$ and $5a^3b^2$, the letters and exponents in both being the same. But in the binomial $8a^2b+7ab^2$, the terms are not similar; for, although they are composed of the same letters, yet the same letters are not affected with like exponents.

24. When an algebraic expression contains similar terms, it may be reduced to a simpler form.

1. Take the expression $3ab+2ab$, which is evidently equal to $5ab$.

2. Reduce the expression $3ac+9ac+2ac$ to its simplest form. *Ans.* $14ac$.

3. Reduce the expression $abc+4abc+5abc$ to its simplest form.

In adding similar terms together we take the sum of the coefficients and annex the literal part. The first term, abc, has a coefficient 1 understood, (Art. **12**).

$$\begin{array}{r} abc \\ 4abc \\ 5abc \\ \hline 10abc \end{array}$$

25. Of the different terms which compose a polynomial, some are preceded by the sign $+$, and the others by the sign $-$. The first are called *additive terms*, the others *subtractive terms.*

QUEST.—**23.** What are similar terms of a polynomial? Are $3a^2b$ and $6a^2b^2$ similar? Are $2a^2b^2$ and $2a^3b^2$? **24.** If the terms are positive and similar, may they be reduced to a simpler form? In what way?

The first term of a polynomial is commonly not preceded by any sign, but then it is understood to be affected with the sign $+$.

1. John has 20 apples and gives 5 to William: how many has he left?

Now, let us represent the number of apples which John has by a, and the number given away by b: the number he would have left would then be represented by $a-b$.

2. A merchant goes into trade with a certain sum of money, say a dollars; at the end of a certain time he has gained b dollars: how much will he then have?

Ans. $a+b$ dollars.

If instead of gaining he had lost b dollars, how much would he have had? *Ans.* $a-b$ dollars.

Now, if the losses exceed the amount with which he began business, that is, if b were greater than a, we must prefix the minus sign to the remainder to show that the quantity to be subtracted was the greatest.

Thus, if he commenced business with \$2000, and lost \$3000, the true difference would be -1000: that is, the subtractive quantity exceeds the additive by \$1000.

3. Let a merchant call the debts due him additive, and the debts he owes subtractive. Now, if he has due him \$600 from one man, \$800 from another, \$300 from another, and owes \$500 to one, \$200 to a second, and \$50 to a third, how will the account stand? *Ans.* \$950 due him.

4. Reduce to its simplest form the expression

$$3a^2b+5a^2b-3a^2b+4a^2b-6a^2b-a^2b,$$

QUEST.—**25.** What are the terms called which are preceded by the sign $+$? What are the terms called which are preceded by the sign $-$ If no sign is prefixed to a term, what sign is understood? If some of the terms are additive and some subtractive, may they be reduced if similar? Give the rule for reducing them. Does the reduction affect the exponents, or only the coefficients?

Additive terms.	*Subtractive terms.*
$+3a^2b$	$-3a^2b$
$+5a^2b$	$-6a^2b$
$+4a^2b$	$-a^2b$
Sum $+12a^2b$	Sum $-10a^2b.$

But, $12a^2b-10a^2b=2a^2b.$

Hence, for the reduction of the similar terms of a polynomial we have the following

RULE.

I. *Add together the coefficients of all the additive terms, and annex to their sum the literal part; and form a single subtractive term in the same manner.*

II. *Then, subtract the less coefficient from the greater, and to the remainder prefix the sign of the greater coefficient, to which annex the literal part.*

REMARK.—It should be observed that the reduction affects only coefficients, and not the exponents.

EXAMPLES.

1. Reduce to its simplest form the polynomial

$$+2a^3bc^2-4a^3bc^2+6a^3bc^2-8a^3bc^2+11a^3bc^2.$$

Find the sum of the additive and subtractive terms separately, and take their difference: thus,

Additive terms.	*Subtractive terms.*
$+2a^3bc^2$	$-4a^3bc^2$
$+6a^3bc^2$	$-8a^3bc^2$
$+11a^3bc^2$	Sum $-12a^3bc^2$
Sum $+19a^3bc^2$	

Hence, the given polynomial reduces to

$$19a^3bc^2-12a^3bc^2=7a^3bc^2$$

2. Reduce the polynomial $4a^2b-8a^2b-9a^2b+11a^2b$ to its simplest form. *Ans.* $-2a^2b$.

3. Reduce the polynomial $7abc^2-abc^2-7abc^2+8abc^2+6abc^2$ to its simplest form. *Ans.* $13abc^2$.

4. Reduce the polynomial $9cb^3-8ac^2+15cb^3+8ca+9ac^2-24cb^3$ to its simplest form. *Ans.* ac^2+8ca.

The reduction of similar terms is an operation peculiar to algebra. Such reductions are constantly made in *Algebraic Addition, Subtraction, Multiplication,* and *Division.*

ADDITION.

26. Addition in Algebra, consists in finding the simplest equivalent expression for several algebraic quantities. Such equivalent expression is called their *sum.*

1. What is the sum of

$$3ax+2ab \text{ and } -2ax+ab.$$

We reduce the terms as in Art. 25, and find for the sum . . . .

$$\begin{array}{r} 3ax+2ab \\ -2ax+\ ab \\ \hline ax+3ab \end{array}$$

2. Let it be required to add together the expressions :

$$\left\{\begin{array}{c} 3a \\ 5b \\ 2c \end{array}\right.$$

The result is . . . . $3a+5b+2c$

an expression which cannot be reduced to a more simple form.

QUEST.—26. What is addition in Algebra? What is such simplest and equivalent expression called?

Again, add together the monomials $\left\{\begin{array}{r} 4a^2b^3 \\ 2a^2b^3 \\ 7a^2b^3 \end{array}\right.$

The result after reducing (Art. **25**), is . . $13a^2b^3$

3. Let it be required to find the sum of the expressions $\left\{\begin{array}{r} 2a^2-4ab \qquad \\ 3a^2-3ab+\ b^2 \\ 2ab-5b^2 \end{array}\right.$

Their sum, after reducing (Art. 25) is . $5a^2-5ab-4b^2$

27. As a course of reasoning similar to the above would apply to all polynomials, we deduce for the addition of algebraic quantities the following general

RULE.

I. *Write down the quantities to be added so that the similar terms shall fall under each other, and give to each term its proper sign.*

II. *Reduce the similar terms, and annex to the results the terms which cannot be reduced, giving to each term its respective sign.*

EXAMPLES.

1. What is the sum of $3ax$, $5ax$, $-2ax$ and $13ax$. ?

Ans. $19ax$.

2. What is the sum of $4ab+8ac$ and $2ab-7ac+d$. ?

Ans. $6ab+ac+d$.

3. Add together the polynomials,

$3a^2-2b^2-4ab$, $5a^2-b^2+2ab$, and $3ab-3c^2-2b^2$.

The term $3a^2$ being similar to $5a^2$, we write $8a^2$ for the result of the reduction of these two terms, at the same time slightly crossing them, as in the first term.

$$\begin{array}{r} 3\not{a}^2-4\not{a}b-2b^2 \qquad \\ 5\not{a}^2+2\not{a}b-\ b^2 \qquad \\ +3\not{a}b-2b^2-3c^2 \\ \hline 8a^2+\ ab-5b^2-3c^2 \end{array}$$

QUEST.—**27**. Give the rule for the addition of Algebraic quantities.

Passing then to the term $-4ab$, which is similar to $+2ab$ and $+3ab$, the three reduce to $+ab$, which is placed after $8a^2$, and the terms crossed like the first term Passing then to the terms involving b^2, we find their sum to be $-5b^2$, after which we write $-3c^2$.

The marks are drawn across the terms, that none of them may be overlooked and omitted.

(4)
$$\begin{array}{r} a \\ a \\ \hline 2a \end{array}$$

(5)
$$\begin{array}{r} 6a \\ 5a \\ \hline 11a \end{array}$$

(6)
$$\begin{array}{r} 5a \\ 5b \\ \hline 5a+5b \end{array}$$

(7)
$$\begin{array}{r} 3ab \\ 5ab \\ \hline 8ab \end{array}$$

(8)
$$\begin{array}{r} 3ac \\ 8ac \\ \hline 11ac \end{array}$$

(9)
$$\begin{array}{r} 7abc+9ax \\ -3abc-3ax \\ \hline 4abc+6ax \end{array}$$

(10)
$$\begin{array}{r} 8ax+3b \\ 5ax-9b \\ \hline 3ax-6b \end{array}$$

(11)
$$\begin{array}{r} 12a-\ 6c \\ -3a-\ 9c \\ \hline 9a-15c \end{array}$$

Note.—If $a=5$, $b=4$, $c=2$, $x=1$, what are the values of the several sums above found.

(12)
$$\begin{array}{r} 9a+f \\ -6a+g \\ -2a-f \\ \hline a+g \end{array}$$

(13)
$$\begin{array}{r} 6ax-\ 8ac \\ -7ax-\ 9ac \\ ax+17ac \\ \hline 0\quad\ 0 \end{array}$$

(14)
$$\begin{array}{r} 3af+\ g\ +m \\ ag-3af-m \\ ab-\ ag+3g \\ \hline ab+4g \end{array}$$

(15)
$$\begin{array}{r} 7x+3ab+\ 3c \\ -3x-3ab-\ 5c \\ 5x-9ab-\ 9c \\ \hline 9x-9ab-11c \end{array}$$

(16)
$$\begin{array}{r} 8x^2+\ 9acx+13a^2b^2c^2 \\ -7x^2-13acx+14a^2b^2c^2 \\ -4x^2+\ 4acx-20a^2b^2c^2 \\ \hline -3x^2+\ 0\quad+\ 7a^2b^2c^2 \end{array}$$

(17)
$$\begin{array}{r} 22h-3c-7f+3g \\ -\ 3h+8c-2f-9g+5x \\ \hline 19h+5c-9f-6g+5x \end{array}$$

(18)
$$\begin{array}{r} 19ah^2+3a^3b^4-8ax^3 \\ -17ah^2-9a^3b^4+9ax^3 \\ \hline 2ah^2-6a^3b^4+\ ax^3 \end{array}$$

(19)	(20)
$7x-9y+5z+3-g$	$8a+b$
$-x-3y \quad -8-g$	$2a-b+c$
$-x+y-3z+1+7g$	$-3a+b \quad +2d$
$-2x+6y+3z-1-g$	$-6b-3c+3d$
$x+8y-5z+9+g$	$-5a \quad +7c-8d$
$4x+3y+0 \;+4+5g$	$2a-5b+5c-3d$

21. Add together $-b+3c-d-115e+6f-5g$, $3b-2c-3d-e+27f$, $5c-8d+3f-7g$, $-7b-6c+17d+9e-5f+11g$, $-3b-5d-2e+6f-9g+h$.

Ans. $-8b-109e+37f-10g+h$.

22. Add together the polynomials $7a^2b-3abc-8b^2c-9c^3+cd^2$, $8abc-5a^2b+3c^3-4b^2c+cd^2$ and $4a^2b-8c^3+9b^2c-3d^3$. *Ans.* $6a^2b+5abc-3b^2c-14c^3+2cd^2-3d^3$.

23. What is the sum of $5a^2bc+6bx-4af$, $-3a^2bc-6bx+14af$, $-af+9bx+2a^2bc$, $+6af-8bx+6a^2bc$.

Ans. $10a^2bc+bx+15af$.

24. What is the sum of $a^2n^2+3a^3m+b$, $-6a^2n^2-6a^3m-b$, $+9b-9a^3m-5a^2n^2$. *Ans.* $-10a^2n^2-12a^3m+9b$.

25. What is the sum of $4a^3b^2c-16a^4x-9ax^3d$, $+6a^3b^2c-6ax^3d+17a^4x$, $+16ax^3d-a^4x-9a^3b^2c$.

Ans. $a^3b^2c+ax^3d$.

26. What is the sum of $-7g+3b+4g-2b$, $+3g-3b+2b$. *Ans.* 0.

27. What is the sum of $ab+3xy-m-n$, $-6xy-3m+11n+cd$, $+3xy+4m-10n+fg$. *Ans.* $ab+cd+fg$.

28. What is the sum of $4xy+n+6ax+9am$, $-6xy+6n-6ax-8am$, $2xy-7n+ax-am$. *Ans.* $+ax$.

29. Add the polynomials $19a^2x^3b-12a^3cb$, $5a^2x^3b+14a^3cb-10ax$, $-2a^2x^3b-12a^3cb$, and $-18a^2x^3b-12a^3cb+9ax$.

Ans. $4a^2x^3b-22a^3cb-ax$.

30. Add together $3a+b+c$, $5a+2b+3ac$, $a+c+ac$, and $-3a-9ac-8b$. *Ans.* $6a-5b+2c-5ac$.

31. Add together $5a^2b+6cx+9bc^2$, $7cx-8a^2b$, and $-15cx-9bc^2+2a^2b$. *Ans.* $-a^2b-2cx$.

32. Add together $8ax+5ab+3a^2b^2c^2$, $-18ax+6a^2+10ab$ and $10ax-15ab-6a^2b^2c^2$. *Ans.* $-3a^2b^2c^2+6a^2$.

33. Add together $3a^2+5a^2b^2c^2-9a^3x$, $7a^2-8a^2b^2c^2-10a^3x$ and $10ab+16a^2b^2c^2+19a^3x$. *Ans.* $10a^2+13a^2b^2c^2+10ab$

SUBTRACTION.

28. Subtraction, in Algebra, consists in finding the simplest expression for the difference between two algebraic quantities.

Thus, the difference between $6a$ and $3a$ is expressed by

$$6a-3a=3a;$$

and the difference between $7a^3b$ and $3a^3b$ by

$$7a^3b-3a^3b=4a^3b.$$

In like manner the difference between $4a$ and $3b$ **is** expressed by $4a-3b$.

Hence, *If the quantities are positive and similar, subtract the coefficients, and to their difference annex the literal part. If they are not similar, place the minus sign before the quantity to be subtracted.*

QUEST.—28. In what does subtraction in Algebra consist? How do you find this difference when the quantities are positive and similar? When they are not similar, how do you express the difference?

	(1)	(2)	(3)
From	$3ab$	$6ax$	$9abc$
take	$2ab$	$3ax$	$7abc$
Rem.	ab	$3ax$	$2abc$.

	(4)	(5)	(6)
From	$16a^2b^2c$	$17a^3b^3c$	$24a^2b^2x$
take	$9a^2b^2c$	$3a^3b^3c$	$7a^2b^2x$
Rem.	$7a^2b^2c$	$14a^3b^3c$	$17a^2b^2x$

	(7)	(8)	(9)
From	$3ax$	$4abx$	$2am$
take	$8c$	$9ac$	ax
Rem.	$3ax-8c$	$4abx-9ac$	$2am-ax$.

29. Let it be required to subtract from $4a$
the binomial . . . . $2b-3c$
The difference may be put under the form $4a-(2b-3c)$.
We must now remark that it is the *difference* between $2b$ and $3c$ which is to be taken from $4a$.

If then, we write . . . . $4a-2b$,
we shall have taken away too much by the units in $3c$; hence, $3c$ must be added to give the true remainder, which is . . . . $4a-2b+3c$.

To illustrate this example by figures, suppose $a=5$, $b=5$, and $c=3$.

We shall then have . . . . $4a=20$
and . . . . $2b-3c=10-9=1$
which may be written . $4a-(2b-3c)=20-1=19$.

QUEST.—29. If $2b-3c$ is to be taken from $4a$, what is proposed to be done? If you subtract $2b$ from $4a$, have you taken too much? How then must you supply the deficiency?

Here it is required to subtract 1 from 20. If, then, we subtract $2b=10$, from $4a=20$, it is plain that we shall have taken too much by $3c=9$, which must therefore be added to give the true remainder.

30. Hence, for the subtraction of algebraic quantities we have the following general

RULE.

I. *Write the quantity to be subtracted under that from which it is to be taken, placing the similar terms, if there are any, under each other.*

II. *Change the signs of all the terms of the polynomial to be subtracted, or conceive them to be changed, and then reduce the polynomial result to its simplest form.*

EXAMPLES.

	(1)	The same with the signs of the lower line changed.	(1)
From	$6ac-5ab+c^2$		$6ac-5ab+c^2$
Take	$3ac+3ab+7c$		$-3ac-3ab-7c$
Rem.	$3ac-8ab+c^2-7c.$		$3ac-8ab+c^2-7c.$

	(2)	(3)
From	$6ax-a+3b^2$	$6yx-3x^2+5b$
Take	$9ax-x+b^2$	$yx-3+a$
Rem.	$-3ax-a+x+2b^2.$	$5yx-3x^2+3+5b-a.$

	(4)	(5)
From	$5a^3-4a^2b+3b^2c$	$4ab-cd+3a^2$
Take	$-2a^3+3a^2b-8b^2c$	$5ab-4cd+3a^2+5b^2$
Rem.	$7a^3-7a^2b+11b^2c.$	$-ab+3cd-5b^2.$

QUEST.—30. Give the rule for the subtraction of algebraic quantities.

6. From $6am+y$ take $3am-x$. *Ans.* $3am+x+y$.

7. From $3ax$ take $3ax-y$. *Ans.* $+y$

8. From $7a^2b^2-x^2$ take $18a^2b^2+x^2$.
Ans. $-11a^2b^2-2x^2$.

9. From $-7f+3m-8x$ take $-6f-5m-2x+3d+8$.
Ans. $-f+8m-6x-3d-8$

10. From $-a-5b+7c-d$ take $4b-c+2d+2k$.
Ans. $-a-9b+8c-3d-2k$

11. From . . $-3a+b-8c+7e-5f+3h-7x-13y$ take $k+2a-9c+8e-7x+7f-y-3l-k$.
Ans. $-5a+b+c-e-12f+3h-12y+3l$

12. From $a+b$ take $a-b$. *Ans.* $2b$.

13. From $2x-4a-2b+5$ take $8-5b+a+6x$.
Ans. $-4x-5a+3b-3$

14. From $3a+b+c-d-10$ take $c+2a-d$.
Ans. $a+b-10$

15. From $3a+b+c-d-10$ take $b-19+3a$.
Ans. $c-d+9$

16. From $2ab+b^2-4c+bc-b$ take $3a^2-c+b^2$.
Ans. $2ab-3a^2-3c+bc-b$.

17. From $a^3+3b^2c+ab^2-abc$ take b^3+ab^2-abc.
Ans. $a^3+3b^2c-b^3$.

18. From $12x+6a-4b+40$ take $4b-3a+4x+6d-10$.
Ans. $8x+9a-8b-6d+50$.

19. From $2x-3a+4b+6c-50$ take $9a+x+6b-6c-40$.
Ans. $x-12a-2b+12c-10$.

20. From $6a-4b-12c+12x$ take $2x-8a+4b-6c$.
Ans. $14a-8b-6c+10x$.

21. From $8abc-12b^3a+6cx-7xy$ take $7cx-xy-13b^3a$.
Ans. $8abc+b^3a-cx-6xy$

31. By the rule for subtraction, polynomials may be subjected to certain transformations.

For example, . . $6a^2 - 3ab + 2b^2 - 2bc$,
becomes . . . . . $6a^2 - (3ab - 2b^2 + 2bc)$.
In like manner . . $7a^3 - 8a^2b - 4b^2c + 6b^2$,
becomes . . . . $7a^3 - (8a^2b + 4b^2c - 6b^2)$,
or, again, ... $7a^3 - 8a^2b - (4b^2c - 6b^2)$.
Also, ... $8a^3 - 7b^3 + c - d$,
becomes . . . . $8a^3 - (7b^2 - c + d)$.
Also, ... $9b^3 - a + 3a^2 - d$,
becomes . . . . $9b^3 - (a - 3a^2 + d)$.

32. Remark.—From what has been shown in addition and subtraction, we deduce the following principles.

1st. In algebra, the words *add* and *sum* do not always, as in arithmetic, convey the idea of augmentation. For, if to a we add $-b$, we have $a-b$, which is, properly speaking, a difference between the number of units expressed by a, and the number of units expressed by b. Consequently, this result is numerically less than a. To distinguish this sum from an arithmetical sum, it is called the *algebraic sum*. p27

Thus, the polynomial $2a^2 - 3a^2b + 3b^2c$ is an algebraic sum, so long as it is considered as the result of the union

---

Quest.—**31**. How may you change the form of a polynomial? **32**. In algebra do the words *add* and *sum* convey the same idea as in arithmetic? What is the algebraic sum of 9 and —4? Of 8 and —2? May an algebraic sum ever be negative? What is the sum of 4 and —8? Do the words subtraction and difference in algebra always convey the idea of diminution? What is the algebraic difference between 8 and —4? Between a and $-b$?

of the monomials $2a^2$, $-3a^2b$, $+3b^2c$, with their respective signs; and, in its *proper acceptation*, it is the arithmetical difference between the sum of the units contained in the additive terms, and the sum of the units contained in the subtractive terms.

It follows from this, that an algebraic sum may, in the numerical applications, be reduced to a *negative* number, or a number affected with the sign $-$.

2nd. The words *subtraction* and *difference* do not always convey the idea of diminution; for, the *numerical difference* between $+a$ and $-b$ being $a+b$, exceeds a. This result is an *algebraic difference*, and can be put under the form of

$$a-(-b)=a+b$$

MULTIPLICATION.

33. If a man earns a dollars in one day, how much will he earn in 6 days? Here it is simply required to repeat the number a, 6 times, which gives $6a$ for the amount earned.

1. What will ten yards of cloth cost at c dollars per yard?
Ans. $10c$ *dollars.*

2. What will d hats cost at 9 dollars per hat?
Ans. $9d$ *dollars.*

3. What will b cravats cost at 40 cents each?
Ans. $40b$ *cents.*

4. What will b pair of gloves cost at a cents a pair?

QUEST.—**33.** What is the object of multiplication in algebra? If a man earns a dollars in one day, how much will he earn in 4 days? In 5 days? In 6 days?

Here it is plain that the cost will be found by repeating b as many times as there are units in a: Hence, the cost is ab cents. *Ans. ab cents.*

NOTE.—If we suppose $a=6$, $c=4$, and $d=3$, what would be the numerical values of the above answers?

5. If a man's income is $3a$ dollars a week, how much will it be in $4b$ weeks. Here we must repeat $3a$ dollars as many times as there are units in $4b$ weeks; hence, the product is equal to

$$3a \times 4b = 12ab.$$

If we suppose $a=4$ and $b=3$ the product will be equal to 144.

34. REMARK.—It is plain that the product $12ab$ will not be altered by changing the arrangement of the factors; that is, $12ab$ is the same as $ab \times 12$, or as $ba \times 12$, or as $a \times 12 \times b$ (See Arithmetic, § **22**).

35. Let us now multiply $3a^2b^2$ by $2a^2b$, which may be placed under the form

$$3a^2b^2 \times 2a^2b = 3 \times 2aaaabbb;$$

in which a is a factor four times, and b a factor three times: hence (Art. **13**).

$$3a^2b^2 \times 2a^2b = 3 \times 2aaaabbb = 6a^4b^3,$$

in which, *we multiply the co-efficients together and add the exponents of the like letters.*

QUEST.—**34**. Will a product be altered by changing the arrangement of the factors? Is $3ab$ the same as $3ba$? Is it the same as $a \times 3b$? As $b \times 3a$? **35**. In multiplying monomials what do you do with the co-efficients? What do you do with the exponents of the common letters? If a letter is found in one factor and not in the other, what do you do?

Hence, for the multiplication of monomials, we have the following

RULE.

I. *Multiply the coefficients together for a new coefficient.*

II. *Write after this coefficient all the letters which enter into the multiplicand and multiplier, affecting each with an exponent equal to the sum of its exponents in both factors.*

EXAMPLES.

1. $$8a^2bc^2 \times 7abd^2 = 56a^3b^2c^2d^2.$$

2. $$21a^3b^2cd \times 8abc^3 = 168a^4b^3c^4d.$$

3. $$4abc \times 7df = 28abcdf.$$

	(4)	(5)	(6)
Multiply	$3a^2b$	$12a^2x$	$6xy\ z$
by	$2a^2b$	$12x^2y$	ay^2z
	$6a^4b^2$	$144a^2x^3y$	$6axy^3z^2.$

(7)	(8)	(9)
a^2xy	$3ab^2c^3$	$87ax^2y$
$2xy^2$	$9a^2b^3c$	$3b^3x^4y^3$
$2a^2x^2y^3$	$27a^3b^5c^4$	$261ab^3x^6y^4.$

10. Multiply $5a^3b^2x^2$ by $6c^5x^6$. *Ans.* $30a^3b^2c^5x^8$

11. Multiply $10a^4b^5c^8$ by $7acd$. *Ans.* $70a^5b^5c^9d$.

12. Multiply $9a^3bxy$ by $9a^3bxy$. *Ans.* $81a^6b^2x^2y^2$.

13. Multiply $36a^8b^7c^6d^5$ by $20ab^2c^3d^4$. *Ans.* $720a^9b^9c^9d^9$

14. Multiply $27axyz$ by $9a^2b^2c^2d^2xyz$.
Ans. $243a^3b^2c^2d^2x^2y^2z^2$

15. Multiply $13a^3b^2c$ by $8abxy$. *Ans* $104a^4b^3cxy$

16. Multiply $20a^5b^5cd$ by $12a^2x^2y$. *Ans.* $240a^7b^5cdx^2y$.

17. Multiply $14a^4b^6d^4y$ by $20a^3c^2x^2y$.
Ans. $280a^7b^6c^2d^4x^2y^2$.

18. Multiply $8a^3b^3y^4$ by $7a^4bxy^5$. *Ans.* $56a^7b^4xy^9$.

19. Multiply $75axyz$ by $5a^5bcdx^2y^2$.
Ans. $375a^6bcdx^3y^3z$.

20. Multiply $51a^2y^2x^2$ by $9a^2bc^2x^5y$. *Ans.* $459a^4bc^2x^7y$.

21. Multiply $2a^3b^2y^2$ by $18abx$. *Ans.* $36a^4b^3xy^3$.

22. Multiply $64a^3m^5x^4yz$ by $8ab^2c^3$.
Ans. $512a^4b^2c^3m^5x^4yz$.

23. Multiply $9a^2b^2c^2d^3$ by $12a^3b^4c^6$. *Ans.* $108a^5b^6c^8d^3$.

24. Multiply $216ab^7c^3d^8$ by $3a^3b^2c^5$. *Ans.* $648a^4b^9c^8d^8$.

25. Multiply $70a^8b^7c^4d^2fx$ by $12a^7b^5c^3dx^2y^3$.
Ans. $840a^{15}b^{12}c^7d^3fx^3y^3$.

36. We will now consider the most general case of two polynomials.

Let a represent the sum of all the additive terms of the multiplicand, and b the sum of the subtractive terms. Let c denote the sum of the additive terms of the multiplier, and d the sum of the subtractive terms. The multiplicand will then be represented by $a-b$, and the multiplier by $c-d$; and it is required to take $a-b$ as many times as there are units in $c-d$.

$$\begin{array}{l} a-b \\ \underline{c-d} \\ ac-bc \\ \quad\;\; \underline{-ad+bd} \\ ac-bc-ad+bd. \end{array}$$

Let us first take $a-b$ as many times as there are units in c.

We begin by writing ac, which is too great by b taken c times; for it is only the *difference* between a and b which is to be taken c times. Hence, $ac-bc$ is the product of $a-b$ by c.

But it was proposed to take $a-b$ only as many times as there are units in the *difference* between c and d: hence the last product $ac-bc$ is too large by $a-b$ taken d times; which without regarding the sign of d, is $ad-bd$. Changing the signs of this last product, and subtracting it from that of $a-b$ by c (Art. **30**), and we have

$$(a-b)\times(c-d)=ac-bc-ad+bd.$$

37. Hence, we have the following rule for the signs.

When two terms of the multiplicand and multiplier are affected with the same sign, the corresponding product is affected with the sign +; and when they are affected with contrary signs, the product is affected with the sign —.

Therefore we say in algebraic language, that + multiplied by +, or — multiplied by —, gives +; — multiplied by +, or + multiplied by —, gives —.

Hence, for the multiplication of polynomials we have the following

RULE.

Multiply all the terms of the multiplicand by each term of the multiplier, observing that like signs give plus in the product, and unlike signs minus. Then reduce the polynomial result to its simplest form.

EXAMPLES IN WHICH ALL THE TERMS ARE PLUS.

1 Multiply . . . . $3a^2+4ab+b^2$

by . . . . $2a+5b$

$6a^3+8a^2b+2ab^2$

The product, after reducing, . $+15a^2b+20ab^2+5b^3$

becomes . . . . . . $6a^3+23a^2b+22ab^2+5b^3$.

QUEST.—**37.** What does + multiplied by + give? + multiplied by —? — multiplied by +? — multiplied by —? Give the rule for the multiplication of polynomials.

2. Multiply $x^2+2ax+a^2$ by $x+a$.
Ans. $x^3+3ax^2+3a^2x+a^3$.

3. Multiply x^3+y^3 by $x+y$. Ans. $x^4+xy^3+x^3y+y^4$.

4. Multiply $3ab^2+6a^2c^2$ by $3ab^2+3a^2c^2$.
Ans. $9a^2b^4+27a^3b^2c^2+18a^4c^4$.

5. Multiply $a^2b^2+c^2d$ by $a+b$.
Ans. $a^3b^2+ac^2d+a^2b^3+bc^2d$.

6. Multiply $3ax^2+9ab^3+cd^5$ by $6a^2c^2$.
Ans. $18a^3c^2x^2+54a^3c^2b^3+6a^2c^3d^5$.

7. Multiply $64a^3x^3+27a^2x+9ab$ by $8a^3cd$.
Ans. $512a^6cdx^3+216a^5cdx+72a^4bcd$.

8. Multiply $a^2+2ax+x^2$ by $a+x$.
Ans. $a^3+3a^2x+3ax^2+x^3$.

9. Multiply $a^3+3a^2x+3ax^2+x^3$ by $a+x$.
Ans. $a^4+4a^3x+6a^2x^2+4ax^3+x^4$.

10. Multiply x^2+y^2 by $x+y$. Ans. $x^3+xy^2+x^2y+y^3$.

11. Multiply x^5+xy^6+7ax by $ax+5ax$.
Ans. $6ax^6+6ax^2y^6+42a^2x^2$.

12. Multiply $a^3+3a^2b+3ab^2+b^3$ by $a+b$.
Ans. $a^4+4a^3b+6a^2b^2+4ab^3+b^4$.

13. Multiply $x^3+x^2y+xy^2+y^3$ by $x+y$.
Ans. $x^4+2x^3y+2x^2y^2+2xy^3+y^4$.

14. Multiply x^3+2x^2+x+3 by $3x+1$.
Ans. $3x^4+7x^3+5x^2+10x+3$

GENERAL EXAMPLES.

1. Multiply	$2ax - 3ab$
by	$3x - b$.
The product	$6ax^2 - 9abx$
becomes after	$- 2abx + 3ab^2$
reducing	$6ax^2 - 11abx + 3ab^2$.

2. Multiply a^4-2b^3 by $a-b$.

Ans. $a^5-2ab^3-a^4b+2b^4$

3. Multiply x^2-3x-7 by $x-2$.

Ans. x^3-5x^2-x+14

4. Multiply $3a^2-5ab+2b^2$ by a^2-7ab.

Ans. $3a^4-26a^3b+37a^2b^2-14ab^3$

5. Multiply $b^2+b^4+b^6$ by b^2-1. *Ans.* b^8-b^2

6. Multiply $x^4-2x^3y+4x^2y^2-8xy^3+16y^4$ by $x+2y$.

Ans. x^5+32y^5.

7. Multiply $4x^2-2y$ by $2y$. *Ans.* $8x^2y-4y^2$.

8. Multiply $2x+4y$ by $2x-4y$. *Ans.* $4x^2-16y^2$.

9. Multiply $x^3+x^2y+xy^2+y^3$ by $x-y$.

Ans. x^4-y^4.

10. Multiply x^2+xy+y^2 by x^2-xy+y^2.

Ans. $x^4+x^2y^2+y^4$

11. Multiply $2a^2-3ax+4x^2$ by $5a^2-6ax-2x^2$.

Ans. $10a^4-27a^3x+34a^2x^2-18ax^3-8x^4$.

12. Multiply $3x^2-2xy+5$ by $x^2+2xy-3$.

Ans. $3x^4+4x^3y-4x^2-4x^2y^2+16xy-15$.

13. Multiply $3x^3+2x^2y^2+3y^2$ by $2x^3-3x^2y^2+5y^3$.

Ans. $\begin{cases} 6x^6-5x^5y^2-6x^4y^4+6x^3y^2+ \\ 15x^3y^3-9x^2y^4+10x^2y^5+15y^5. \end{cases}$

14. Multiply $8ax-6ab-c$ by $2ax+ab+c$.

Ans. $16a^2x^2-4a^2bx-6a^2b^2+6acx-7abc-c^2$.

15. Multiply $3a^2-5b^2+3c^2$ by a^2-b^2.

Ans. $3a^4-8a^2b^2+3a^2c^2+5b^4-3b^2c^2$.

16.

$$\begin{array}{l} 3a^2-5bd+cf \\ -5a^2+4bd-8cf. \\ \hline \end{array}$$

Pro. red. $-15a^4+37a^2bd-29a^2cf-20b^2d^2+44bcdf-8c^2f^2$.

38. To finish with what has reference to algebraic multiplication, we will make known a few results of frequent use in Algebra.

Let it be required to form the square or second power of the binomial $(a+b)$. We have, from known principles,

$$(a+b)^2=(a+b)\ (a+b)=a^2+2ab+b^2.$$

That is, *the square of the sum of two quantities is equal to the square of the first, plus twice the product of the first by the second, plus the square of the second.*

1. Form the square of $2a+3b$. We have from the rule

$$(2a+3b)^2 = 4a^2 + 12ab + 9b^2.$$

2. $$(5ab+3ac)^2 =25a^2b^2+ 30a^2bc+ 9a^2c^2.$$

3. $$(5a^2+8a^2b)^2 =25a^4 + 80a^4b +64a^4b^2.$$

4. $$(6ax+9a^2x^2)^2=36a^2x^2+108a^3x^3+81a^4x^4.$$

39. To form the square of a difference $a-b$, we have

$$(a-b)^2=(a-b)\ (a-b)=a^2-2ab+b^2.$$

That is, *the square of the difference between two quantities is equal to the square of the first, minus twice the product of the first by the second, plus the square of the second.*

1. Form the square of $2a-b$. We have

$$(2a-b)^2=4a^2-4ab+b^2.$$

2. Form the square of $4ac-bc$. We have

$$(4ac-bc)^2=16a^2c^2-8abc^2+b^2c^2.$$

3. Form the square of $7a^2b^2-12ab^3$. We have

$$(7a^2b^2-12ab^3)^2=49a^4b^4-168a^3b^5+144a^2b^6.$$

QUEST.—**38**. What is the square of the sum of two quantities equal to? **39**. What is the square of the difference of two quantities equal to?

40. Let it be required to multiply $a+b$ by $a-b$. We have

$$(a+b)\times(a-b)=a^2-b^2.$$

Hence, *the sum of two quantities, multiplied by their difference, is equal to the difference of their squares*

1. Multiply $2c+b$ by $2c-b$. We have

$$(2c+b)\times(2c-b)=4c^2-b^2.$$

2. Multiply $9ac+3bc$ by $9ac-3bc$. We have

$$(9ac+3bc)(9ac-3bc)=81a^2c^2-9b^2c^2.$$

3. Multiply $8a^3+7ab^2$ by $8a^3-7ab^2$. We have

$$(8a^3+7ab^2)(8a^3-7ab^2)=64a^6-49a^2b^4.$$

41. It is sometimes convenient to find the factors of a polynomial, or to resolve a polynomial into its factors. Thus, if we have the polynomial

$$ac+ab+ad,$$

we see that a is a common factor to each of the terms: hence, it may be placed under the form

$$a(c+b+d).$$

1. Find the factors of the polynomial $a^2b^2+a^2d+a^2f$.

Ans. $a^2(b^2+d+f)$.

2. Find the factors of $3a^2b+6a^2b^2+b^2d$.

Ans. $b(3a^2+6a^2b+bd)$.

3. Find the factors of $3a^2b+9a^2c+18a^2xy$.

Ans. $3a^2(b+3c+6xy)$.

QUEST.—**40.** What is the sum of two quantities multiplied by their difference equal to?

4. Find the factors of $8a^2cx-18acx^2+2ac^5y-30a^6c^9x$.

Ans. $2ac(4ax-9x^2+c^4y-15a^5c^8x)$.

5. Find the factors of $a^2+2ab+b^2$.

Ans. $(a+b)\times(a+b)$.

6. Find the factors of a^2-b^2. *Ans.* $(a+b)\times(a-b)$.

7. Find the factors of $a^2-2ab+b^2$.

Ans. $(a-b)\times(a-b)$.

DIVISION.

42. Algebraic division has the same object as arithmetical, viz: having given a product, and one of its factors, to find the other factor.

We will first consider the case of two monomials.

The division of $72a^5$ by $8a^3$ is indicated thus:

$$\frac{72a^5}{8a^3}.$$

It is required to find a third monomial, which, multiplied by the second, will produce the first. It is plain that the third monomial is $9a^2$; for by the rules of multiplication

$$8a^3\times9a^2=72a^5.$$

Hence, we have $$\frac{72a^5}{8a^3}=9a^2,$$

a result which is obtained by *dividing the coefficient of the dividend by the coefficient of the divisor, and subtracting the exponents of the like letter.*

QUEST.—**42.** What is the object of division in Algebra? Give the rule for dividing monomials?

Also, $\frac{35a^3b^2c}{7ab}=5a^{3-1}b^{2-1}c=5a^2bc,$

for, $7ab\times 5a^2bc=35a^3b^2c.$

Again, $\frac{56a^4b^2c^2}{8a^3bc}=7abc.$

Hence, for the division of monomials we have the following

RULE.

I. *Divide the coefficient of the dividend by the coefficient of the divisor, for a new coefficient.*

II. *Write after this coefficient, all the letters of the dividend, and affect each with an exponent equal to the excess of its exponent in the dividend over that in the divisor.*

From these rules we find

$$\frac{48a^3b^3c^2d}{12ab^2c}=4a^2bcd;\quad \frac{150a^5b^8cd^3}{30a^3b^5d^2}=5a^2b^3cd.$$

1. Divide $16x^2$ by $8x$. *Ans.* $2x$.
2. Divide $15ax^2y^3$ by $3ay$. *Ans.* $5x^2y^2$.
3. Divide $84ab^3x$ by $12b^2$. *Ans.* $7abx$.
4. Divide $36a^4b^5c^2$ by $9a^3b^2c$. *Ans.* $4ab^3c$.
5. Divide $88a^3b^2c$ by $8a^2b$. *Ans.* $11abc$.
6. Divide $99a^4b^4x^5$ by $11a^3b^2x^4$. *Ans.* $9ab^2x$
7. Divide $108x^6y^5z^3$ by $54x^5z$. *Ans.* $2xy^5z^2$.
8. Divide $64x^7y^5z^6$ by $16x^6y^4z^5$. *Ans.* $4xyz$.
9. Divide $96a^7b^6c^5$ by $12a^2bc$. *Ans.* $8a^5b^5c^4$.
10. Divide $54a^7c^5d^6$ by $27acd$. *Ans.* $2a^6c^4d^5$.
11. Divide $38a^4b^6d^4$ by $2a^3b^5d$. *Ans.* $19abd^3$.

12. Divide $42a^2b^2c^2$ by $7abc$. *Ans.* $6abc$.
13. Divide $64a^5b^4c^8$ by $32a^4bc$. *Ans.* $2ab^3c^7$.
14. Divide $128a^5x^6y^7$ by $16axy^4$. *Ans.* $8a^4x^5y^3$.
15. Divide $132bd^5f^6$ by $2d^4f$. *Ans.* $66bdf^5$.
16. Divide $256a^4b^9c^8d^7$ by $16a^3bc^6$. *Ans.* $16ab^8c^2d^7$
17. Divide $200a^8m^2n^2$ by $50a^7mn$. *Ans.* $4amn$.
18. Divide $300x^3y^4z^2$ by $60xy^2z$. *Ans.* $5x^2y^2z$.
19. Divide $27a^5b^2c^2$ by $9abc$. *Ans.* $3a^4bc$.
20. Divide $64a^3y^6z^8$ by $32ay^5z^7$. *Ans.* $2a^2yz$.
21. Divide $88a^5b^6c^8$ by $11a^3b^4c^6$. *Ans.* $8a^2b^2c^2$.

43. It follows from the preceding rule, that the division of monomials will be impossible,

1st. When the coefficients are not divisible by each other.

2nd. When the exponent of the same letter is greater in the divisor than in the dividend.

3rd. When the divisor contains one or more letters which are not found in the dividend.

When either of these three cases occurs, the quotient remains under the form of a monomial fraction; that is, a monomial expression, necessarily containing the algebraic sign of division, but which may frequently be reduced.

Take, for example, $12a^4b^2cd$, to be divided by $8a^2bc^2$, which is placed under the form

$$\frac{12a^4b^2cd}{8a^2bc^2};$$

QUEST.—**43.** What is the first case named in which the division of monomials will not be exact? What is the second? What is the third? If either of these cases occur, can the exact division be made? Under what form will the quotient then remain? Mav this fraction be often reduced to a simpler form?

this may be reduced by dividing the numerator and denominator by the common factors 4, a^2, b, and c, which gives

$$\frac{12a^4b^2cd}{8a^2bc^2}=\frac{3a^2bd}{2c}.$$

Also,
$$\frac{25a^5b^2d^3}{15a^4b^6d^4}=\frac{5a}{3b^4d}.$$

44. Hence, for the reduction of a monomial fraction to its simplest form, we have the following

RULE.

Suppress every factor, whether numerical or literal, that is common to both terms of the fraction, and the result will be the reduced fraction sought.

From this new rule, we find,

(1)
$$\frac{48a^3b^2cd^3}{36a^2b^3c^2de}=\frac{4ad^2}{3bce}:$$
and (2)
$$\frac{37a\,b^3c^5d}{6a^3b\,c^4d^2}=\frac{37b^2c}{6a^2d};$$

also (3)
$$\frac{7a^2b}{14a^3b^2}=\frac{1}{2ab}:$$
and (4)
$$\frac{4a^2b^2}{6ab^4}=\frac{2a}{3b^2}.$$

5. Divide $49a^2b^2c^6$ by $14a^3bc^4$. *Ans.* $\frac{7bc^2}{2a}$

6. Divide $6amn$ by $3abc$. *Ans.* $\frac{2mn}{bc}$.

7. Divide $18a^2b^2mn^2$ by $12a^4b^4cd$. *Ans.* $\frac{3mn^2}{2a^2b^2cd}$

QUEST.—**44.** Give the rule for the reduction of a monomial fraction.

8. Divide $28a^5b^6c^7d^8$ by $16ab^9cd^7m$. *Ans.* $\frac{7a^4c^6d}{4b^3m}$.

9 Divide $72a^3c^2b^2$ by $12a^5c^4b^3d$. *Ans.* $\frac{6}{a^2c^2bd}$.

10. Divide $100a^8b^5xmn$ by $25a^3b^4d$. *Ans.* $\frac{4a^5bxmn}{d}$.

11. Divide $96a^5b^8c^9df$ by $75a^2cxy$. *Ans.* $\frac{32a^3b^8c^8df}{25xy}$.

12. Divide $85m^2n^3fx^2y^3$ by $15am^4nf$. *Ans.* $\frac{17n^2x^2y^3}{3am^2}$.

13. Divide $127d^3x^2y^2$ by $16d^4x^4y^4$. *Ans.* $\frac{127}{16dx^2y^2}$.

45. If we have an expression of the form

$$\frac{a}{a},\quad \text{or}\quad \frac{a^2}{a^2},\quad \text{or}\quad \frac{a^3}{a^3},\quad \text{or}\quad \frac{a^4}{a^4},\quad \text{or}\quad \frac{a^5}{a^5},\quad \&c,$$

and apply the rule for the exponents, we shall have

$$\frac{a}{a}=a^{1-1}=a^0,\quad \frac{a^2}{a^2}=a^{2-2}=a^0,\quad \frac{a^3}{a^3}=a^{3-3}=a^0,\quad \&c.$$

But since any quantity divided by itself is equal to 1, it follows that

$$\frac{a}{a}=a^0=1,\quad \frac{a^2}{a^2}=a^{2-2}=a^0=1,\quad \&c,$$

or finally, if we designate the general exponent by m, we have

$$\frac{a^m}{a^m}=a^{m-m}=a^0=1\ ;$$

that is, *any power of which the exponent is* 0 *is equal to* 1; and hence, a *factor* of the form a^0, being equal to 1, may be omitted.

QUEST.—**45.** What is a^0 equal to? What is b^0 equal to? **What is** the power of *any number* equal to, when the **exponent of the power is** 0?

2. Divide $6a^2b^2c^4d$ by $2a^2b^2d$.

$$\frac{6a^2b^2c^4d}{2a^2b^2d}=3a^{2-2}b^{2-2}c^4d^{1-1}=3c^4.$$

3. Divide $8a^4b^3c^4d^5$ by $4a^2b^3c^4d^5$. *Ans.* $2a^2$

4. Divide $16a^6b^8d^9$ by $8a^6b^8d$. *Ans.* $2d^8$.

5. Divide $32m^3n^3x^2y^2$ by $4m^3n^3xy$. *Ans.* $8xy$.

6. Divide $96a^4b^5d^8c^9$ by $24a^4b^4d^5c^9$. *Ans.* $4bd^3$.

SIGNS IN DIVISION.

46. The object of division, is to find a third quantity called the quotient, which, multiplied by the divisor, shall produce the dividend.

Since, in multiplication, the product of two terms having the same sign is affected with the sign +, and the product of two terms having contrary signs is affected with the sign −, we may conclude,

1st. That when the term of the dividend has the sign +, and that of the divisor the sign of +, the term of the quotient must have the sign +.

2nd. When the term of the dividend has the sign +, and that of the divisor the sign −, the term of the quotient must have the sign −, because it is only the sign −, which, multiplied with the sign −, can produce the sign + of the dividend.

QUEST.—**46.** What will the quotient, multiplied by the divisor, be equal to? If the multiplicand and multiplier have like signs, what will be the sign of the product? If they have contrary signs, what will be the sign of the product? When the term of the dividend and the term of the divisor have the same sign, what will be the sign of the quotient? When they have different signs, what will be the sign of the quotient?

3rd. When the term of the dividend has the sign —, and that of the divisor the sign +, the quotient must have the sign —. Again we say for brevity, that,

+ divided by +, and — divided by —, give + ;
— divided by +, and + divided by —, give —.

EXAMPLES.

1. Divide $4ax$ by $-2a$. *Ans.* $-2x$.

Here it is plain that the answer must be $-2x$; for,

$-2a \times -2x = +4ax$, the divisor

2. Divide $36a^3x^2$ by $-12a^2x$. *Ans.* $-3ax$.
3. Divide $-58a^3b^5c^2d^2$ by $29a^2b^4c$. *Ans.* $-2abcd^2$.
4. Divide $-84a^4b^5d^3$ by $-42a^2b^2d$. *Ans.* $2a^2b^3d^2$.
5. Divide $64c^4d^5x^3$ by $16c^4dx$. *Ans.* $4d^4x^2$.
6. Divide $-88b^4x^5y^6$ by $-24b^3cdx^5$. *Ans.* $+\frac{11by^6}{3cd}$.
7. Divide $77a^4y^3z^4$ by $-11a^4y^3z^4$. *Ans.* -7.
8. Divide $84a^4b^2c^2d$ by $-42a^4b^2c^2d$. *Ans.* -2.
9. Divide $-60a^7b^6c^4d$ by $-12a^8b^7c^5d^2$. *Ans.* $+5\frac{1}{abcd}$.
10. Divide $-88a^6b^7c^6$ by $8a^5b^6c^6$. *Ans.* $-11ab$.
11. Divide $16x^2$ by $-8x$. *Ans.* $-2x$.
12. Divide $-15a^2xy^3$ by $3ay$. *Ans.* $-5axy^2$.
13. Divide $-84ab^3x$ by $-12b^2$. *Ans.* $7abx$.
14. Divide $-96a^4b^2c^3$ by $12a^3bc$. *Ans.* $-8abc^2$.
15. Divide $-144a^9b^8c^7d^5$ by $-36a^4b^6c^6d$. *Ans.* $4a^5b^2cd^4$.
16. Divide $256a^3bc^2x^3$ by $-16a^2cx^2$. *Ans.* $-16abcx$.
17. Divide $-300a^5b^4c^3x^2$ by $30a^4b^3c^2x$. *Ans.* $-10abcx$.
18 Divide $500a^8b^9c^6$ by $-100a^7b^8c^4$. *Ans.* $-5abc^2$.

19. Divide $-64a^5b^8c^7$ by $-8a^4b^7c^6$. *Ans.* $8abc$.
20. Divide $+96a^5b^4d^9$ by $-24a^4b^2d$. *Ans.* $-4ab^2d^8$.
21. Divide $72a^5b^3d^4$ by $-8a^4b^2d$. *Ans.* $-9abd^3$.

Division of Polynomials.

FIRST EXAMPLE.

47. Divide $a^2-2ax+x^2$ by $a-x$.

It is found most convenient, in division in algebra, to place the divisor on the right of the dividend, and the quotient directly under the divisor.

Dividend.	*Divisor.*
$a^2-2ax+x^2$	$a-x$
a^2-ax	$a-x$ *Quotient.*
$-ax+x^2$	
$-ax+x^2$	

We first divide the term a^2 of the dividend by the term a of the divisor: the partial quotient is a, which we place under the divisor. We then multiply the divisor by a, and subtract the product a^2-ax from the dividend, and to the remainder bring down x^2. We then divide the first term of the remainder, $-ax$ by a, the quotient is $-x$. We then multiply the divisor by $-x$, and, subtracting as before, we find nothing remains. Hence, $a-x$ is the exact quotient.

In this example, *we have written the terms of the dividend and divisor in such a manner that the exponents of the same letter shall go on diminishing from left to right.* This is what is called *arranging* the dividend and divisor with reference to a certain letter. By this preparation, the first term on the left of the dividend, and the first on the left of the divisor, are always the two which must be divided by each other in order to obtain a term of the quotient.

QUEST.—**47.** What do you understand by arranging a polynomial with reference to a particular letter?

48. Hence, for the division of polynomials we have the following

RULE.

I. *Arrange the dividend and divisor with reference to a certain letter, and then divide the first term on the left of the dividend by the first term on the left of the divisor, the result is the first term of the quotient; multiply the divisor by this term, and subtract the product from the dividend.*

II. *Then divide the first term of the remainder by the first term of the divisor, which gives the second term of the quotient; multiply the divisor by the second term, and subtract the product from the result of the first operation. Continue the same process until you obtain 0 for a remainder; in which case the division is said to be exact.*

SECOND EXAMPLE.

Let it be required to divide

$$51a^2b^2+10a^4-48a^3b-15b^4+4ab^3 \text{ by } 4ab-5a^2+3b^2.$$

We here arrange with reference to a.

$$\begin{array}{l|l}
\textit{Dividend.} & \textit{Divisor.} \\
10a^4-48a^3b+51a^2b^2+\ 4ab^3-15b^4 & -5a^2+4ab+3b^2 \\
\underline{+10a^4-\ 8a^3b-\ 6a^2b^2} & \overline{-2a^2+8ab-5b^2} \\
\quad -40a^3b+57a^2b^2+\ 4ab^3-15b^4 & \quad \textit{Quotient.} \\
\quad \underline{-40a^3b+32a^2b^2+24ab^3} & \\
\qquad\quad 25a^2b^2-20ab^3-15b^4 & \\
\qquad\quad \underline{25a^2b^2-20ab^3-15b^4} &
\end{array}$$

QUEST.—48. Give the general rule for the division of polynomials? If the first term of the arranged dividend is not divisible by the first term of the arranged divisor, is the exact division possible? If the first term of any partial dividend is not divisible by the first term of the divisor, is the exact division possible?

REMARK.—When the first term of the arranged dividend is not exactly divisible by that of the arranged divisor, the complete division is impossible ; that is to say, there is not a polynomial which, multiplied by the divisor, will produce the dividend. And in general, we shall find that a division is impossible, when the first term of one of the partial dividends is not divisible by the first term of the divisor.

GENERAL EXAMPLES.

1. Divide $18x^2$ by $9x$. *Ans.* $2x$.
2. Divide $10x^2y^2$ by $-5x^2y$. *Ans.* $-2y$.
3. Divide $-9ax^2y^2$ by $9x^2y$. *Ans.* $-ay$.
4. Divide $-8x^2$ by $-2x$. *Ans.* $+4x$.
5. Divide $10ab+15ac$ by $5a$. *Ans.* $2b+3c$.
6. Divide $30ax-54x$ by $6x$. *Ans.* $5a-9$.
7. Divide $10x^2y-15y^2-5y$ by $5y$. *Ans.* $2x^2-3y-1$.
8. Divide $12a+3ax-18ax^2$ by $3a$. *Ans.* $4+x-6x^2$.
9. Divide $6ax^2+9a^2x+a^2x^2$ by ax. *Ans.* $6x+9a+ax$.
10. Divide $a^2+2ax+x^2$ by $a+x$. *Ans.* $a+x$.
11. Divide $a^3-3a^2y+3ay^2-y^3$ by $a-y$.
Ans. $a^2-2ay+y^2$.
12. Divide $24a^2b-12a^3cb^2-6ab$ by $-6ab$.
Ans. $-4a+2a^2cb+1$.
13. Divide $6x^4-96$ by $3x-6$. *Ans.* $2x^3+4x^2+8x+16$
14. Divide . . . $a^5-5a^4x+10a^3x^2-10a^2x^3+5ax^4-x^5$ by $a^2-2ax+x^2$. *Ans.* $a^3-3a^2x+3ax^2-x^3$.
15. Divide $48x^3-76ax^2-64a^2x+105a^3$ by $2x-3a$.
Ans. $24x^2-2ax-35a^2$.

16. Divide $y^6-3y^4x^2+3y^2x^4-x^6$ by $y^3-3y^2x+3yx^2-x^3$.
Ans. $y^3+3y^2x+3yx^2+x^3$.

17. Divide $64a^4b^6-25a^2b^8$ by $8a^2b^3+5ab^4$.
Ans. $8a^2b^3-5ab^4$.

18. Divide $6a^3+23a^2b+22ab^2+5b^3$ by $3a^2+4ab+b^2$.
Ans. $2a+5b$.

19. Divide $6ax^6+6ax^2y^6+42a^2x^2$ by $ax+5ax$.
Ans. x^5+xy^6+7ax.

20. Divide . . $-15a^4+37a^2bd-29a^2cf-20b^2d^2+44bcdf$ $8c^2f^2$ by $3a^2-5bd+cf$. *Ans.* $-5a^2+4bd-8cf$.

21. Divide $x^4+x^2y^2+y^4$ by x^2-xy+y^2.
Ans. x^2+xy+y^2.

22. Divide x^4-y^4 by $x-y$. *Ans.* $x^3+x^2y+xy^2+y^3$.

23. Divide $3a^4-8a^2b^2+3a^2c^2+5b^4-3b^2c^2$ by a^2-b^2.
Ans. $3a^2-5b^2+3c^2$.

24. Divide . . $6x^6-5x^5y^2-6x^4y^4+6x^3y^2+15x^3y^3-9x^2y^4$ $+10x^2y^5+15y^5$ by $3x^3+2x^2y^2+3y^2$.
Ans. $2x^3-3x^2y^2+5y^3$.

25. Divide . $-c^2+16a^2x^2-7abc-4a^2bx-6a^2b^2+6acx$ by $8ax-6ab-c$. *Ans.* $2ax+ab+c$

26. Divide . . . . $3x^4+4x^3y-4x^2-4x^2y^2+16xy-15$ by $2xy+x^2-3$. *Ans.* $3x^2-2xy+5$.

27. Divide x^5+32y^5 by $x+2y$.
Ans. $x^4-2x^3y+4x^2y^2-8xy^3+16y^4$.

28. Divide $3a^4-26a^3b-14ab^3+37a^2b^2$ by $2b^2-5ab$ $+3a^2$. *Ans.* a^2-7ab.

CHAPTER II.

Algebraic Fractions.

49. Algebraic fractions should be considered in the same point of view as arithmetical fractions, such as $\frac{3}{4}$, $\frac{11}{12}$; that is, we must conceive that the unit has been divided into as many equal parts as there are units in the denominator, and that one of these parts is taken as many times as there are units in the numerator. Hence, addition, subtraction, multiplication, and division, are performed according to the rules established for arithmetical fractions.

It will not, therefore, be necessary to demonstrate those rules, and in their application we must follow the procedures indicated for the operations on entire algebraic quantities.

50. Every quantity which is not expressed under a fractional form is called an *entire* algebraic quantity.

51. An algebraic expression, composed partly of an entire quantity and partly of a fraction, is called a *mixed quantity.*

QUEST.—**49** How are algebraic fractions to be considered? What does the denominator show? What does the numerator show? How then are the operations in fractions to be performed? **50**. What is an entire quantity? **51**. What is a mixed quantity?

CASE I.

To reduce a fraction to its simplest terms.

52. A fraction is said to be in its lowest terms, when there is no common factor in the numerator and denominator. The rule for reducing a monomial fraction to its lowest terms has already been given (Art. **44**).

With respect to polynomial fractions, the following are cases which are easily reduced.

1. Take, for example, the expression

$$\frac{a^2-b^2}{a^2-2ab+b^2}.$$

This fraction can take the form

$$\frac{(a+b)\,(a-b)}{(a-b)^2},$$

(Art. **39** and **40**). Suppressing the factor $a-b$, which is common to the two terms, we obtain

$$\frac{a+b}{a-b}.$$

2. Again, take the expression

$$\frac{5a^3-10a^2b+5ab^2}{8a^3-8a^2b}.$$

This expression can be decomposed thus:

$$\frac{5a(a^2-2ab+b^2)}{8a^2(a-b)},$$

or,
$$\frac{5a\,(a-b)^2}{8a^2(a-b)}.$$

QUEST.—52. How do you reduce a fraction to its simplest terms?

Suppressing the common factors $a(a-b)$, the result is

$$\frac{5(a-b)}{8a}.$$

Hence, *to reduce any fraction to its simplest terms, we suppress or cancel every factor common to the numerator and denominator.*

NOTE.—Find the common factors of the numerator and denominator as explained in (Art. **41**).

EXAMPLES.

1. Reduce $\frac{3a^2+6a^2b^2}{12a^4+6a^3c^2}$ to its simplest terms.

Ans. $\frac{1+2b^2}{4a^2+2ac^2}$.

2. Reduce $\frac{15a^5c+25a^9d}{25a^2+30a^2}$ to its simplest terms.

Ans. $\frac{3a^3c+5a^7d}{11}$.

3. Reduce $\frac{85b^7cd^5}{15b^7c^8d^5}$ to its simplest terms.

Ans. $\frac{17}{3c^7}$

4. Reduce $\frac{60c^6d^4f^5}{12c^5d^8f^9}$ to its simplest terms.

Ans. $\frac{5c}{d^4f^4}$.

5. Reduce $\frac{27a^4b^4-81a\,b^6}{63a\,b^6-36a^2b^4}$. Ans. $\frac{3a^3-9b^2}{7b^2-4a}$

6. Reduce $\frac{96a^3b^2c}{-12a^3b^2c}$ to its simplest terms. Ans. -8.

7. Reduce $\frac{24b^5-36ab^4}{48a^4b^4-66a^5b^6}$. Ans. $\frac{4b-6a}{8a^4-11a^5b^2}$.

CASE II.

53. To reduce a mixed quantity to the form of a fraction

RULE.

Multiply the entire part by the denominator of the fraction; then connect this product with the terms of the numerator by the rules for addition, and under the result place the given denominator.

EXAMPLES.

1. Reduce $6\frac{1}{7}$ to the form of a fraction

$$6\times7=42: \quad 42+1=43: \quad \text{hence,} \quad 6\tfrac{1}{7}=\frac{43}{7}.$$

2. Reduce $x-\frac{(a^2-x^2)}{x}$ to the form of a fraction.

$$x-\frac{a^2-x^2}{x}=\frac{x^2-(a^2-x^2)}{x}=\frac{2x^2-a^2}{x}. \quad \textit{Ans.}$$

3. Reduce $x-\frac{ax+x^2}{2a}$ to the form of a fraction.

Ans. $\frac{ax-x^2}{2a}$.

4. Reduce $5+\frac{2x-7}{3x}$ to the form of a fraction.

Ans. $\frac{17x-7}{3x}$.

5. Reduce $1-\frac{x-a-1}{a}$ to the form of a fraction.

Ans. $\frac{2a-x+1}{a}$

QUEST.—**53.** How do you reduce a mixed quantity to the form of a fraction?

6. Reduce $1+2x-\frac{x-3}{5x}$ to the form of a fraction.

Ans. $\frac{10x^2+4x+3}{5x}$.

7. Reduce $2a+b-\frac{3c+4}{8}$ to the form of a fraction.

Ans. $\frac{16a+8b-3c-4}{8}$

8. Reduce $6ax+b-\frac{6a^2x-ab}{4a}$ to the form of a fraction.

Ans. $\frac{18a^2x+5ab}{4a}$.

9. Reduce $8+3ab-\frac{8+6a^2b^2x^4}{12abx^4}$ to the form of a fraction.

Ans. $\frac{96abx^4+30a^2b^2x^4-8}{12abx^4}$.

10 Reduce $9+\frac{3b^2-8c^4}{a-b^2}$ to the form of a fraction.

Ans. $\frac{9a-6b^2-8c^4}{a-b^2}$.

CASE III.

54. To reduce a fraction to an entire or mixed quantity.

RULE.

Divide the numerator by the denominator for the entire part, and place the remainder, if any, over the denominator for the fractional part.

QUEST.—**54**. How do you reduce a fraction to an entire or mixed quantity?

EXAMPLES.

1. Reduce $\frac{8966}{8}$ to an entire number.

$$\begin{array}{l} 8)8966(\\ \hline 1120 \ldots 6 \text{ rem.} \end{array}$$

Hence, $1120\frac{6}{8}=$ *Ans.*

2. Reduce $\frac{ax-a^2}{x}$ to a mixed quantity.

Ans. $a-\frac{a^2}{x}$.

3. Reduce $\frac{ax-x^2}{x}$ to an entire or mixed quantity.

Ans. $a-x$.

4. Reduce $\frac{ab-2a^2}{b}$ to a mixed quantity.

Ans. $a-\frac{2a^2}{b}$.

5. Reduce $\frac{a^2-x^2}{a-x}$ to an entire quantity. *Ans.* $a+x$.

6. Reduce $\frac{x^3-y^3}{x-y}$ to an entire quantity.

Ans. x^2+xy+y^2.

7 Reduce $\frac{10x^2-5x+3}{5x}$ to a mixed quantity.

Ans. $2x-1+\frac{3}{5x}$.

8. Reduce $\frac{36x^3-72x+32a^2x^2}{9x}$ to a mixed quantity.

Ans. $4x^2-8+\frac{32a^2x}{9}$.

CASE IV.

55. To reduce fractions having different denominators to equivalent fractions having a common denominator.

RULE.

Multiply each numerator into all the denominators except its own, for the new numerators, and all the denominators together for a common denominator.

EXAMPLES.

1. Reduce $\frac{1}{2}$, $\frac{7}{3}$, and $\frac{4}{5}$, to a common denominator.

	$1\times3\times5=15$	the new numerator of the 1st.
	$7\times2\times5=70$	" " " 2nd.
	$4\times3\times2=24$	" " " 3rd.
and	$2\times3\times5=30$	the common denominator.

Therefore, $\frac{15}{30}$, $\frac{70}{30}$, and $\frac{24}{30}$, are the equivalent fractions.

NOTE.—It is plain that this reduction does not alter the values of the several fractions, since the numerator and denominator of each are multiplied by the same number.

2. Reduce $\frac{a}{b}$ and $\frac{b}{c}$ to equivalent fractions having a common denominator.

$\left.\begin{array}{l} a\times c=ac \\ b\times b=b^2 \end{array}\right\}$ the new numerators.

and $b\times c=bc$ the common denominator.

QUEST.—55. How do you reduce fractions to a common denominator?

Hence, $\frac{ac}{bc}$ and $\frac{b^2}{bc}$ are the equivalent fractions.

3. Reduce $\frac{a}{b}$ and $\frac{a+b}{c}$ to fractions having a common denominator. *Ans.* $\frac{ac}{bc}$ and $\frac{ab+b^2}{bc}$.

4. Reduce $\frac{3x}{2a}$, $\frac{2b}{3c}$, and d, to fractions having a common denominator. *Ans.* $\frac{9cx}{6ac}$, $\frac{4ab}{6ac}$, and $\frac{6acd}{6ac}$.

5. Reduce $\frac{3}{4}$, $\frac{2x}{3}$, and $a+\frac{2x}{a}$, to fractions having a common denominator.

Ans. $\frac{9a}{12a}$, $\frac{8ax}{12a}$, and $\frac{12a^2+24x}{12a}$.

6. Reduce $\frac{1}{2}$, $\frac{a^2}{3}$, and $\frac{a^2+x^2}{a+x}$, to fractions having a common denominator.

Ans. $\frac{3a+3x}{6a+6x}$, $\frac{2a^3+2a^2x}{6a+6x}$, and $\frac{6a^2+6x^2}{6a+6x}$.

7. Reduce $\frac{a}{3b}$, $\frac{6ax}{5c}$, and $\frac{a^2-x^2}{d}$ to a common denominator.

Ans. $\frac{5acd}{15bcd}$, $\frac{18abdx}{15bcd}$, and $\frac{15a^2bc-15bcx^2}{15bcd}$

8. Reduce $\frac{c}{5a}$, $\frac{a-b}{c}$, and $\frac{c}{a+b}$, to a common denominator.

Ans. $\frac{ac^2+c^2b}{5a^2c+5abc}$, $\frac{5a^3-5ab^2}{5a^2c+5abc}$, and $\frac{5ac^2}{5a^2c+5abc}$.

CASE V.

56. To add fractional quantities together

RULE.

Reduce the fractions, if necessary, to a common denominator; then add the numerators together, and place their sum over the common denominator.

EXAMPLES.

1. Add $\frac{6}{2}$, $\frac{4}{3}$, and $\frac{2}{5}$ together.

By reducing to a common denominator, we have

$6\times3\times5=90$ 1st numerator.
$4\times2\times5=40$ 2nd numerator.
$2\times3\times2=12$ 3rd numerator.
$2\times3\times5=30$ the denominator.

Hence, the fractions become

$$\frac{90}{30}+\frac{40}{30}+\frac{12}{30}=\frac{142}{30};$$

which, by reducing to the lowest terms become $4\frac{11}{15}$.

2. Find the sum of $\frac{a}{b}$, $\frac{c}{d}$, and $\frac{e}{f}$.

Here $a\times d\times f=adf$, $c\times b\times f=cbf$, $e\times b\times d=ebd$ } the new numerators

And $b\times d\times f=bdf$ the common denominator.

Hence, $\frac{adf}{bdf}+\frac{cbf}{bdf}+\frac{ebd}{bdf}=\frac{adf+cbf+ebd}{bdf}$ the sum.

QUEST.—**56.** How do you add fractions.

3 To $a-\frac{3x^2}{b}$ add $b+\frac{2ax}{c}$.

Ans. $a+b+\frac{2abx-3cx^2}{bc}$.

4. Add $\frac{x}{2}$, $\frac{x}{3}$ and $\frac{x}{4}$ together. Ans. $x+\frac{x}{12}$.

5. Add $\frac{x-2}{3}$ and $\frac{4x}{7}$ together. Ans. $\frac{19x-14}{21}$.

6. Add $x+\frac{x-2}{3}$ to $3x+\frac{2x-3}{4}$. Ans. $4x+\frac{10x-17}{12}$.

7. It is required to add $4x$, $\frac{5x^2}{2a}$, and $\frac{x+a}{2x}$ together.

Ans. $4x+\frac{5x^3+ax+a^2}{2ax}$.

8. It is required to add $\frac{2x}{3}$, $\frac{7x}{4}$, and $\frac{2x+1}{5}$ together.

Ans. $2x+\frac{49x+12}{60}$.

9. It is required to add $4x$, $\frac{7x}{9}$, and $2+\frac{x}{5}$ together.

Ans. $4x+\frac{44x+90}{45}$.

10. It is required to add $3x+\frac{2x}{5}$ and $x-\frac{8x}{9}$ together.

Ans. $3x+\frac{23x}{45}$.

11. Required the sum of $ac-\frac{6b}{8a}$ and $1-\frac{c}{d}$.

Ans. $\frac{8a^2cd-6bd+8ad-8ac}{8ad}$

CASE VI.

57. To subtract one fractional quantity from another

RULE.

I. *Reduce the fractions to a common denominator.*

II. *Subtract the numerator of the fraction to be subtracted from the numerator of the other fraction, and place the difference over the common denominator.*

EXAMPLES.

1. What is the difference between $\frac{3}{7}$ and $\frac{2}{8}$.

$$\frac{3}{7}-\frac{2}{8}=\frac{24}{56}-\frac{14}{56}=\frac{10}{56}=\frac{5}{28}. \quad Ans.$$

2. Find the difference of the fractions $\frac{x-a}{2b}$ and $\frac{2a-4x}{3c}$.

Here, $\left.\begin{array}{l}(x-\ a)\times 3c=3cx-3ac \\ (2a-4x)\times 2b=4ab-8bx\end{array}\right\}$ the numerators

And, $2b\times 3c=6bc$ the common denominator.

Hence, $\frac{3cx-3ac}{6bc}-\frac{4ab-8bx}{6bc}=\frac{3cx-3ac-4ab+8bx}{6bc}$. *Ans*

3. Required the difference of $\frac{12x}{7}$ and $\frac{3x}{5}$. *Ans.* $\frac{39x}{35}$.

4. Required the difference of $5y$ and $\frac{3y}{8}$. *Ans.* $\frac{37y}{8}$

5. Required the difference of $\frac{3x}{7}$ and $\frac{2x}{9}$. *Ans.* $\frac{13x}{63}$.

QUEST.—57. How do you subtract fractions?

6. Required the difference between $\frac{x+a}{b}$ and $\frac{c}{d}$.

Ans. $\frac{dx+ad-bc}{bd}$.

7. Required the difference of $\frac{3x+a}{5b}$ and $\frac{2x+7}{8}$.

Ans. $\frac{24x+8a-10bx-35b}{40b}$.

8. Required the difference of $3x+\frac{x}{b}$ and $x-\frac{x-a}{c}$.

Ans. $2x+\frac{cx+bx-ab}{bc}$.

CASE VII.

58. To multiply fractional quantities together.

RULE.

If the quantities to be multiplied are mixed, reduce them to fractional forms ; then multiply the numerators together for a numerator and the denominators together for a denominator

EXAMPLES.

1. Multiply $\frac{1}{6}$ of $\frac{3}{7}$ by $8\frac{1}{3}$.

Operation.

We first reduce the compound fraction to the simple one $\frac{3}{42}$, and then the mixed number to the equivalent fraction $\frac{25}{3}$; after which, we multiply the numerators and denominators together.

$$\frac{1}{6} \text{ of } \frac{3}{7}=\frac{3}{42},$$

$$8\tfrac{1}{3}=\frac{25}{3}.$$

Hence, $\frac{3}{42}\times\frac{25}{3}=\frac{75}{126}=\frac{25}{42}$

Ans. $\frac{25}{42}$

2. Multiply $a+\frac{bx}{a}$ by $\frac{c}{d}$. First, $a+\frac{bx}{a}=\frac{a^2+bx}{a}$

Hence, $\frac{a^2+bx}{a}\times\frac{c}{d}=\frac{a^2c+bcx}{ad}$. *Ans.*

3. Required the product of $\frac{3x}{5}$ and $\frac{3a}{b}$. *Ans.* $\frac{9ax}{5b}$.

4. Required the product of $\frac{2x}{5}$ and $\frac{3x^2}{2a}$.

Ans. $\frac{3x^3}{5a}$.

5. Find the continued product of $\frac{2x}{a}$, $\frac{3ab}{c}$ and $\frac{3ac}{2b}$.

Ans. $9ax$.

6. It is required to find the product of $b+\frac{bx}{a}$ and $\frac{a}{x}$.

Ans. $\frac{ab+bx}{x}$.

7. Required the product of $\frac{x^2-b^2}{bc}$ and $\frac{x^2+b^2}{b+c}$.

Ans. $\frac{x^4-b^4}{b^2c+bc^2}$.

8. Required the product of $x+\frac{x+1}{a}$, and $\frac{x-1}{a+b}$.

Ans. $\frac{ax^2-ax+x^2-1}{a^2+ab}$.

9. Required the product of $a+\frac{ax}{a-x}$ by $\frac{a^2-x^2}{x+x^2}$.

Ans. $\frac{a^4-a^2x^2}{ax+ax^2-x^2-x^3}$

QUEST.—58. How do you multiply fractions together?

CASE VIII.

59. To divide one fractional quantity by another.

RULE.

Reduce the mixed quantities, if there are any, to fractional forms; then invert the terms of the divisor, and multiply the fractions together as in the last case.

EXAMPLES.

1. Divide . . . . $\frac{10}{24}$ by $\frac{5}{8}$.

Operation.

$$\frac{5}{8}=5\times\frac{1}{8}$$

$$\frac{10}{24}\div5=\frac{10}{120}$$

$$\frac{10}{120}\times8=\frac{80}{120}=\frac{2}{3}.$$

If the divisor were 5, the quotient would be $\frac{10}{120}$. But, since the divisor is $\frac{1}{8}$ of 5, the true quotient must be 8 times $\frac{10}{120}$, for the eighth of a number will be contained in the dividend 8 times more than the number itself. In this operation we have actually multiplied the numerator of the dividend by 8 and the denominator by 5; that is, *we have inverted the terms of the divisor and multiplied the fractions together.*

2. Divide . . $a-\frac{b}{2c}$ by $\frac{f}{g}$.

$$a-\frac{b}{2c}=\frac{2ac-b}{2c}.$$

Hence, $a-\frac{b}{2c}\div\frac{f}{g}=\frac{2ac-b}{2c}\times\frac{g}{f}=\frac{2acg-bg}{2cf}$. *Ans.*

QUEST.—59. How do you divide one fraction by another?

3. Let $\frac{7x}{5}$ be divided by $\frac{12}{13}$. *Ans* $\frac{91x}{60}$.

4. Let $\frac{4x^2}{7}$ be divided by $5x$. *Ans.* $\frac{4x}{35}$.

5. Let $\frac{x+1}{6}$ be divided by $\frac{2x}{3}$. *Ans.* $\frac{x+1}{4x}$.

6. Let $\frac{x}{x-1}$ be divided by $\frac{x}{2}$. *Ans.* $\frac{2}{x-1}$.

7. Let $\frac{5x}{3}$ be divided by $\frac{2a}{3b}$. *Ans.* $\frac{5bx}{2a}$.

8. Let $\frac{x-b}{8cd}$ be divided by $\frac{3cx}{4d}$. *Ans.* $\frac{x-b}{6c^2x}$.

9. Let $\frac{x^4-b^4}{x^2-2bx+b^2}$ be divided by $\frac{x^2+bx}{x-b}$.

Ans. $x+\frac{b^2}{x}$.

10. Divide $6a^2+\frac{b}{5}$ by $c^2-\frac{x-a}{2}$.

Ans. $\frac{60a^2+2b}{10c^2-5x+5a}$.

11. Divide $18c^2-x+\frac{a}{b}$ by $a^2-\frac{b}{5}$.

Ans. $\frac{90bc^2-5bx+5a}{5a^2b-b^2}$.

12. Divide $20x^2-\frac{8ab}{dc^3}$ by $x^2-\frac{b-c}{f}$.

Ans. $\frac{20dc^3fx^2-8abf}{dc^3fx^2-dc^3b+dc^4}$.

CHAPTER III.

Of Equations of the First Degree.

60. An *Equation* is the algebraic expression of two equal quantities with the sign of equality placed between them. Thus, $x=a+b$ is an equation, in which x is equal to the sum of a and b.

61. By the definition, every equation is composed of two parts, separated from each other by the sign $=$. The part on the left of the sign, is called the *first member;* and the part on the right, is called the *second member.* Each member may be composed of one or more terms. Thus, in the equation $x=a+b$, x is the first member, and $a+b$ the second.

62. Every equation may be regarded as the enunciation, in algebraic language, of a particular question. Thus, the equation $x+x=30$, is the algebraic enunciation of the following question:

QUEST.—**60.** What is an equation? **61.** Of how many parts is every equation composed? How are the parts separated from each other? What is the part on the left called? What is the part on the right called? May each member be composed of one or more terms? In the equation $x=a+b$, which is the first member? Which the second? How many terms in the first member? How many in the second?

To find a number which being added to itself, shall give a sum equal to 30.

Were it required to solve this question, we should first express it in algebraic language, which would give the equation

$$x+x=30.$$

By adding x to itself, we have

$$2x=30.$$

And by dividing by 2, we obtain

$$x=15.$$

Hence we see that the solution of a question by algebra consists of two distinct parts: viz. the STATEMENT and the SOLUTION of an equation.

I. The STATEMENT *consists in expressing algebraically the relation between the known and unknown quantities.*

II. The SOLUTION *of the equation consists in finding the value of the unknown quantity in terms of those which are known.*

The given or known parts of a question, are represented either by numbers or by the first letters of the alphabet, a, b, c, &c. The unknown or required parts are represented by the final letters, x, y, z, &c.

EXAMPLE.

Find a number which, being added to twice itself, the sum shall be equal to 24.

QUEST.—**62.** How may you regard every equation? What question does the equation $x+x=30$ state? Of how many parts does the solution of a question by algebra consist? Name them. What is the 2nd part called? By what are the known parts of a question represented? By what are the unknown parts represented?

Statement.

Let x represent the number. We shall then have

$$x+2x=24.$$

This is the statement.

Solution.

Having . . . . $x+2x=24,$
we add . . . . . $x+2x,$
which gives . . . $3x=24,$
and dividing by 3, . $x=8.$

63. An equation is said to be *verified* when the answer found, being substituted for the unknown quantity, proves the two members of the equation to be equal to each other.

Thus, in the last equation we found $x=8$. If we substitute this value for x in the equation

$$x+2x=24,$$

we shall have $8+2\times 8=8+16=24.$

which proves that 8 is the true answer.

64. An equation involving only the first power of the unknown quantity, is called an equation of the *first degree.*

Thus, $6x+3x-5=13,$

and $ax+bx+c=d,$

are equations of the first degree.

By considering the nature of an equation, we perceive that it must possess the three following properties:

QUEST.—**63.** When is an equation said to be *verified*? **64.** When an equation involves only the first power of the unknown quantity what is it called? What are the three properties of every equation?

1st. The two members are composed of quantities of the same kind: that is, dollars=dollars, pounds=pounds, &c.

2nd. The two members are equal to each other.

3rd. The two members must have the same sign.

65. An axiom is a self-evident truth. We may here state the following.

1. *If equal quantities be added to both members of an equation, the equality of the members will not be destroyed.*

2. *If equal quantities be subtracted from both members of an equation, the equality will not be destroyed.*

3. *If both members of an equation be multiplied by the same number, the equality will not be destroyed.*

4. *If both members of an equation be divided by the same number, the equality will not be destroyed.*

Transformation of Equations.

66. The *transformation* of an equation consists in changing its form without affecting the equality of its members.

The following transformations are of continual use in the resolution of equations.

First Transformation.

67. When some of the terms of an equation are fractional, to reduce the equation to one in which the terms shall be entire.

1. Take the equation

$$\frac{2x}{3}-\frac{3}{4}x+\frac{x}{6}=11.$$

QUEST.—**65**. What is an axiom? Name the four axioms. **66**. What is the transformation of an equation? **67**. What is the first transformation? What is the least common multiple of several numbers? How do you find the least common multiple?

First, reduce all the fractions to the same denominator by the known rule; the equation then becomes

$$\frac{48x}{72}-\frac{54x}{72}+\frac{12x}{72}=11;$$

and since we can multiply both members by the same number without destroying the equality, we will multiply them by 72, which is the same as suppressing the denominator 72, in the fractional terms, and multiplying the entire term by 72; the equation then becomes

$$48x-54x+12x=792,$$

or dividing by 6 $\quad 8x-\ 9x+\ 2x=132.$

But this last equation can be obtained in a shorter way, by finding the least common multiple of the denominators.

The least common multiple of several numbers is the least number which they will separately divide without a remainder. When the numbers are small, it may at once be determined by inspection. The manner of finding the least common multiple is fully shown in Arithmetic § 87

Take for example, the last equation

$$\frac{2x}{3}-\frac{3}{4}x+\frac{x}{6}=11.$$

We see that 12 is the least common multiple of the denominators, and if we multiply all the terms of the equation by 12, and divide by the denominators, we obtain

$$8x-9x+2x=132.$$

the same equation as before found.

68. Hence, to make the denominators disappear from an equation, we have the following

RULE.

I. *Find the least common multiple of all the denominators.*

II. *Multiply every term of the equation by the common multiple—reducing at the same time the fractional to entire terms.*

EXAMPLES.

1. Clear the equation of $\frac{x}{5}+\frac{x}{7}-4=3$ of its denominators.
Ans. $7x+5x-140=105$.

2. Clear the equation $\frac{x}{6}+\frac{x}{9}-\frac{x}{27}=8$ of its denominators.
Ans. $9x+6x-2x=432$

3. Clear the equation $\frac{x}{2}+\frac{x}{3}-\frac{x}{9}+\frac{x}{12}=20$ of its denominators.
Ans. $18x+12x-4x+3x=720$.

4. Clear the equation $\frac{x}{5}+\frac{x}{7}-\frac{x}{2}=4$ of its denominators.
Ans. $14x+10x-35x=280$.

5. Clear the equation $\frac{x}{4}-\frac{x}{5}+\frac{x}{6}=15$ of its denominators
Ans. $15x-12x+10x=90C$

QUEST.—68. Give the rule for clearing an equation of its denominators.

6. Clear the equation $\frac{x}{4}-\frac{x}{6}+\frac{x}{8}+\frac{x}{9}=12$ of its denominators.

Ans. $18x-12x+9x+8x=864$

7. Clear the equation $\frac{a}{b}-\frac{c}{d}+f=g$.

Ans. $ad-bc+bdf=bdg$.

8. In the equation

$$\frac{ax}{b}-\frac{2c^2x}{ab}+4a=\frac{4bc^2x}{a^3}-\frac{5a^3}{b^2}+\frac{2c^2}{a}-3b,$$

the least common multiple of the denominators is a^3b^2; hence clearing the fractions, we obtain

$$a^4bx-2a^2bc^2x+4a^4b^2=4b^3c^2x-5a^6+2a^2b^2c^2-3a^3b^3.$$

Second Transformation.

69. When the two members of an equation are entire polynomials, to transpose certain terms from one member to the other.

1. Take for example the equation

$$5x-6=8+2x.$$

If, in the first place we subtract $2x$ from both members, the equality will not be destroyed, and we have

$$5x-6-2x=8.$$

Whence we see that the term $2x$, which was additive in the second member becomes subtractive in the first.

QUEST.—**69.** What is the second transformation? What do you understand by transposing a term? Give the rule for transposing from one member to the other.

In the second place, if we add 6 to both members, the equality will still exist, and we have

$$5x-6-2x+6=8+6.$$

Or, since -6 and $+6$ destroy each other, we have

$$5x-2x=8+6.$$

Hence the term which was subtractive in the first member, passes into the second member with the sign of addition.

2. Again, take the equation

$$ax+l=d-cx.$$

If we add cx to both members and subtract b from them, the equation becomes

$$ax+b+cx-b=d-cx+cx-b.$$

or reducing $$ax+cx=d-b.$$

When a term is taken from one member of an equation and placed in the other, it is said to be *transposed.*

Therefore, for the transposition of the terms, we have the following

RULE.

Any term of an equation may be transposed from one member to the other by changing its sign.

70. We will now apply the preceding principles to the resolution of equations.

1. Take the equation

$$4x-3=2x+5.$$

By transposing the terms -3 and $2x$, it becomes

$$4x-2x=5+3.$$

Or, reducing $$2x=8.$$

Dividing by 2 $$x=\frac{8}{2}=4.$$

Verification.

If now, 4 be substituted in the place of x in the given equation

$$4x-3=2x+5,$$

it becomes $$4\times 4-3=2\times 4+5.$$

or, $$13=13.$$

Hence, the value of x is verified by substituting it for the unknown quantity in the given equation.

2. For a second example, take the equation

$$\frac{5x}{12}-\frac{4x}{3}-13=\frac{7}{8}-\frac{13x}{6}.$$

By making the denominators disappear, we have

$$10x-32x-312=21-52x,$$

or, by transposing

$$10x-32x+52x=21+312$$

by reducing $$30x=333$$

$$x=\frac{333}{30}=\frac{111}{10}=11,1.$$

a result which may be verified by substituting it for x in the given equation.

3. For a third example let us take the equation

$$(3a-x)(c-b)+2ax=4b(x+a).$$

It is first necessary to perform the multiplications indicated, in order to reduce the two members to two polynomials, and thus be able to disengage the unknown quantity x, from the known quantities. Having done that, the equation becomes,

$$3a^2-ax-3ab+bx+2ax=4bx+4ab,$$

or, by transposing

$$-ax+bx+2ax-4bx=4ab+3ab-3a^2,$$

by reducing $\qquad ax-3bx=7ab-3a^2.$

Or, (Art. **41**). $\qquad (a-3b)x=7ab-3a^2.$

Dividing both members by $a-3b$ we find

$$x=\frac{7ab-3a^2}{a-3b}.$$

Hence, in order to resolve an equation of the first degree, we have the following general

RULE.

I. *If there are any denominators, cause them to disappear, and perform, in both members, all the algebraic operations indicated.*

II. *Then transpose all the terms affected with the unknown quantity into the first member, and all the known terms into the second member.*

III. *Reduce to a single term all the terms involving* x: *this term will be composed of two factors, one of which will be* x, *and the other all the multipliers of* x, *connected with their respective signs.*

IV. *Divide both members of the equation by the multiplier of the unknown quantity.*

QUEST.—70. What is the first step in resolving an equation of the first degree? What the second? What the third? What the fourth?

EXAMPLES.

1. Given $3x-2+24=31$ to find x. *Ans.* $x=3$.

2 Given $x+18=3x-5$ to find x. *Ans.* $x=11\frac{1}{2}$.

3. Given $6-2x+10=20-3x-2$ to find x.
Ans. $x=2$.

4. Given $x+\frac{1}{2}x+\frac{1}{3}x=11$ to find x. *Ans.* $x=6$.

5. Given $2x-\frac{1}{2}x+1=5x-2$ to find x. *Ans.* $x=\frac{6}{7}$.

6. Given $3ax+\frac{a}{2}-3=bx-a$ to find x.
Ans. $x=\frac{6-3a}{6a-2b}$.

7. Given $\frac{x-3}{2}+\frac{x}{3}=20-\frac{x-19}{2}$ to find x.
Ans. $x=23\frac{1}{4}$.

8. Given $\frac{x+3}{2}+\frac{x}{3}=4-\frac{x-5}{4}$ to find x.
Ans. $x=3\frac{6}{13}$.

9. Given $\frac{x}{4}-\frac{3x}{2}+x=\frac{4x}{8}-3$ to find x.
Ans. $x=4$.

10. Given $\frac{3ax}{c}-\frac{2bx}{d}-4=f$ to find x.
Ans. $x=\frac{cdf+4cd}{3ad-2bc}$.

11. Given $\frac{8ax-b}{7}-\frac{3b-c}{2}=4-b$ to find x.

Ans. $x=\frac{56+9b-7c}{16a}$.

12. Given $\frac{x}{5}-\frac{x-2}{3}+\frac{x}{2}=\frac{13}{3}$ to find x.

Ans. $x=10$.

13. Given $\frac{x}{a}-\frac{x}{b}+\frac{x}{c}-\frac{x}{d}=f$ to find x.

Ans. $x=\frac{abcdf}{bcd-acd+abd-abc}$.

NOTE.—What is the numerical value of x, when $a=1$, $b=2$, $c=3$, $d=4$, $b=5$, and $f=6$.

14. Given $\frac{x}{7}-\frac{8x}{9}-\frac{x-3}{5}=-12\frac{29}{45}$ to find x.

Ans. $x=14$.

15. Given $x-\frac{3x-5}{13}+\frac{4x-2}{11}=x+1$ to find x.

Ans. $x=6$.

16. Given $x+\frac{x}{4}+\frac{x}{5}-\frac{x}{6}=2x-43$ to find x.

Ans. $x=60$

17. Given $2x-\frac{4x-2}{5}=\frac{3x-1}{2}$ to find x.

Ans. $x=3$.

18 Given $3x+\frac{bx-d}{3}=x+a$ to find x.

Ans. $x=\frac{3a+d}{6+b}$.

19. Given $\frac{ax-b}{4}+\frac{a}{3}=\frac{bx}{2}-\frac{bx-a}{3}$ to find x.

Ans. $x=\frac{3b}{3a-2b}$.

20. Find the value of x in the equation

$$\frac{(a+b)(x-b)}{a-b}-3a=\frac{4ab-b^2}{a+b}-2x+\frac{a^2-bx}{b}.$$

$$Ans.\ x=\frac{a^4+3a^3b+4a^2b^2-6ab^3+2b^4}{2b(2a^2+ab-b^2)}.$$

Of Questions producing Equations of the First Degree involving but a single unknown quantity.

71. It has already been observed (Art. **62**), that the solution of a question by algebra consists of two distinct parts:

1st. To make the STATEMENT: that is, to express the conditions of the question algebraically;

2d. To solve the equation: that is, to disengage the known from the unknown quantities.

We have already explained the manner of finding the value of the unknown quantity, after the question has been stated; and it only remains to point out the best methods of putting a question in the language of algebra.

This part of the algebraic solution of a question cannot, like the second, be subjected to any well defined rule. Sometimes the enunciation of the question furnishes the equation immediately; and sometimes it is necessary to discover, from the enunciation, new conditions from which an equation may be formed.

QUEST.—**71**. Into how many parts is the resolution of a question in algebra divided? What is the first step? What the second? Which part has already been explained? Which part is now to be considered? Can this part be subjected to exact rules? Give the general rule for stating a question

In almost all cases, however, we are able to make the statement; that is, to discover the equation, by applying the following

RULE.

Represent the unknown quantity by one of the final letters of the alphabet; and then indicate by means of the algebraic signs, the same operations on the known and unknown quantities, as would verify the value of the unknown quantity, were such value known.

QUESTIONS.

1. To find a number to which if 5 be added, the sum will be equal to 9.

Denote the number by x.

Then by the conditions

$$x+5=9.$$

This is the statement of the question.

To find the value of x, we transpose 5 to the second member, which gives

$$x=9-5=4.$$

Verification.

$$4+5=9.$$

2. Find a number such, that the sum of one half, one third, and one fourth of it, augmented by 45, shall be equal to 448.

Let the required number be denoted by x.

Then, one half of it will be denoted by $\frac{x}{2}$.

one third " " by $\frac{x}{3}$.

one fourth " " by $\frac{x}{4}$.

And by the conditions,

$$\frac{x}{2}+\frac{x}{3}+\frac{x}{4}+45=448.$$

This is the statement of the question.

To find the value of x, subtract 45 from both members: this gives

$$\frac{x}{2}+\frac{x}{3}+\frac{x}{4}=403.$$

By clearing the terms of their denominators, we obtain

$$6x+4x+3x=4836,$$

or

$$13x=4836.$$

Hence,

$$x=\frac{4836}{13}=372.$$

Verification.

$$\frac{372}{2}+\frac{372}{3}+\frac{372}{4}+45=186+124+93+45=448.$$

3. What number is that whose third part exceeds its fourth by 16.

Let the required number be represented by x. Then,

$\frac{1}{3}x=$ the third part.

$\frac{1}{4}x=$ the fourth part.

And by the question

$$\frac{1}{3}x-\frac{1}{4}x=16.$$

This is the statement. To find the value of x, we clear the terms of the denominators, which gives

$$4x-3x=192.$$

and

$$x=192.$$

Verification.

$$\frac{192}{3}-\frac{192}{4}=64-48=16.$$

4. Divide \$1000 between A, B and C, so that A shall have \$72 more than B, and C \$100 more than A.

Let $x=$ B's share of the \$1000.

Then $x+72=$ A's share,

and $x+172=$ C's share,

their sum is $3x+244=\$1000$.

This is the statement.

By transposing 244 we have

$$3x=1000-244=756$$

and $x=\frac{756}{3}=252=$ B's share.

Hence, $x+72=252+72=\$324=$ A's share.

And $x+172=252+172=\$424=$ C's share.

Verification.

$$252+324+424=1000.$$

5. Out of a cask of wine which had leaked away a third part, 21 gallons were afterwards drawn, and the cask being then guaged, appeared to be half full: how much did it hold?

Suppose the cask to have held x gallons.

Then, $\frac{x}{3}$ what leaked away.

And $\frac{x}{3}+21=$ what had leaked and been drawn

Hence, $\frac{x}{3}+21=\frac{1}{2}x$ by the question.

This is the statement.

To find x. we have

$$2x+126=3x,$$

and $$-x=-126,$$

or $$x=126,$$

by changing the signs of both members, which does not destroy their equality.

Verification.

$$\frac{126}{3}+21=42+21=63=\frac{126}{2}.$$

6. A fish was caught whose tail weighed 9*lb.*, his head weighed as much as his tail and half his body, and his body weighed as much as his head and tail together; what was the weight of the fish?

Let $2x=$ the weight of the body.

Then, $9+x=$ weight of the head;

and since the body weighed as much as both head and tail,

$$2x=9+9+x,$$

which is the statement. Then,

$$2x-x=18 \quad \text{and} \quad x=18.$$

Verification.

$2x=36lb.=$ weight of the body.
$9+x=27lb.=$ weight of the head.
$9lb.=$ weight of the tail.
Hence, $72lb.=$ weight of the fish.

7. The sum of two numbers is 67 and their difference 19: what are the two numbers?

Let $x=$ the least number.

Then, $x+19=$ the greater.

and by the conditions of the question

$$2x+19=67.$$

This is the statement.

To find x, we first transpose 19, which gives.

$$2x=67-19=48;$$

hence, $x=\frac{48}{2}=24$, and $x+19=43$.

Verification.

$$43+24=67, \quad \text{and} \quad 43-24=19.$$

Another Solution.

Let x represent the greater number:

then, $x-19$ will represent the least,

and, $2x-19=67$, whence $2x=67+19$;

therefore, $x=\frac{86}{2}=43$

and consequently $x-19=43-19=24$.

General Solution of this Problem.

The sum of two numbers is a, their difference is b. What are the two numbers?

Let x be the least number,

$x+b$ will represent the greater.

Hence, $2x+b=a$, whence $2x=a-b$;

therefore, $$x=\frac{a-b}{2}=\frac{a}{2}-\frac{b}{2},$$

and consequently, $$x+b=\frac{a}{2}-\frac{b}{2}+b=\frac{a}{2}+\frac{b}{2}.$$

As the form of these two results is independent of any value attributed to the letters a and b, it follows that,

Knowing the sum and difference of two numbers, we obtain the greater by adding the half difference to the half sum, and the less, by subtracting the half difference from half the sum.

Thus, if the given sum were 237, and the difference 99,

the greater is $\frac{237}{2}+\frac{99}{2}$, or $\frac{237+99}{2}=\frac{336}{2}=168$;

and the least $\frac{237}{2}-\frac{99}{2}$, or $\frac{138}{2}=69$.

Verification.

$$168+69=237 \quad \text{and} \quad 168-69=99.$$

8. A person engaged a workman for 48 days. For each day that he laboured he received 24 cents, and for each day that he was idle, he paid 12 cents for his board. At the end of the 48 days, the account was settled, when the labourer received 504 cents. *Required the number of working days, and the number of days he was idle.*

If these two numbers were known, by multiplying them respectively by 24 and 12, then subtracting the last product from the first, the result would be 504. Let us indicate these operations by means of algebraic signs.

Let $x =$ the number of working days.

$48-x =$ the number of idle days.

Then, $24 \times x =$ the amount earned,

and $12(48-x)=$ the amount paid for his board.

Then, $24x-12(48-x)=504$

what he received, which is the statement. Then to find we first multiply by 12, which gives

$$24x-576+12x=504.$$

or,
$$36x=504+576=1080,$$

$$x=\frac{1080}{36}=30 \text{ the working days.}$$

whence,
$$48-30=18 \text{ the idle days.}$$

Verification.

Thirty day's labour, at 24 cents a day, amounts to	$30 \times 24 = 720$ cents
And 18 day's board, at 12 cents a day, amounts to	$18 \times 12 = 216$ cents.
The difference is the amount received	504 cents.

General Solution.

This question may be made general, by denoting the whole number of working and idle days by n.

The amount received for each day he worked by a.

The amount paid for his board, for each idle day, by b

And the balance due the laborer, or the result of the account, by c.

As before, let the number of working days be represented by x.

The number of idle days will be expressed by $n-x$.

Hence, what he earns will be expressed by ax.

And the sum to be deducted, on account of his board, by $b(n-x)$.

The equation of the problem therefore is

$$ax-b(n-x)=c,$$

which is the statement. To find x we first multiply by b, which gives

$$ax-bn+bx=c$$

or,

$$(a+b)x=c+bn$$

whence,

$$x=\frac{c+bn}{a+b}= \text{ working days.}$$

and consequently, $n-x=n-\dfrac{c+bn}{a+b}=\dfrac{an+bn-c-bn}{a+b}$,

or

$$n-x=\frac{an-c}{a+b}= \text{ idle days.}$$

Let us now suppose $n=48$, $a=24$, $b=12$, and $c=504$ These numbers will give for x the same value as before found.

9. A person dying leaves half of his property to his wite, one-sixth to each of two daughters, one-twelfth to a servant and the remaining $600 to the poor: what was the amount of his property?

Represent the amount of the property by x.

Then, $\frac{x}{2}=$ what he left to his wife,

$\frac{x}{6}=$ what he left to one daughter,

and $\frac{2x}{6}=\frac{x}{3}$ what he left to both daughters,

$\frac{x}{12}=$ what he left to his servant.

\$600 to the poor.

Then, by the conditions of the question

$$\frac{x}{2}+\frac{x}{3}+\frac{x}{12}+600=x$$

the amount of the property, which gives $x=\$7200$.

10. A and B play together at cards. A sits down with \$84 and B with \$48. Each loses and wins in turn, when it appears that A has five times as much as B. How much did A win?

Let x represent what A won.

Then A rose with $\$84+x$ dollars,

and B rose with $\$48-x$ dollars.

But by the conditions of the question, we have

$$84+x=5(48-x),$$

hence, $84+x=240-5x$;

and, $6x=156$,

consequently, $x=\$26$ what A won.

Verification.

$$84+26=110;\quad 48-26=22;$$

$$110=5(22)=110$$

11. A can do a piece of work alone in 10 days, B in 13 days: in what time can they do it if they work together?

Denote the time by x, and the work to be done by 1. Then in 1 day A could do $\frac{1}{10}$ of the work, and B could do $\frac{1}{13}$; and in x days A could do $\frac{x}{10}$ of the work, and B, $\frac{x}{13}$: hence, by the conditions of the question

$$\frac{x}{10}+\frac{x}{13}=1,$$

which gives $$13x+10x=130:$$

hence, $$23x=130, \quad x=\frac{130}{23}=5\tfrac{15}{23} \text{ days.}$$

12. A fox, pursued by a greyhound, has a start of 60 leaps. He makes 9 leaps while the greyhound makes but 6; but, three leaps of the greyhound are equivalent to 7 of the fox. How many leaps must the greyhound make to overtake the fox?

From the enunciation, it is evident that the distance to be passed over by the greyhound is composed of the 60 leaps which the fox is in advance, plus the distance that the fox passes over from the moment when the greyhound starts in pursuit of him. Hence, if we can find the expression for these two distances, it will be easy to form the equation of the problem.

Let $x=$ the number of leaps made by the greyhound before he overtakes the fox.

Now, since the fox makes 9 leaps while the greyhound makes but 6, the fox will make $\frac{9}{6}$ or $\frac{3}{2}$ leaps while the greyhound makes 1 ; and, therefore, while the greyhound makes x leaps, the fox will make $\frac{3}{2}x$ leaps.

Hence, the distance which the greyhound must pass over will be expressed by $60+\frac{3}{2}x$ leaps of the fox.

It might be supposed, that in order to obtain the equation, it would be sufficient to place x equal to $60+\frac{3}{2}x$; but in doing so, a manifest error would be committed; for the leaps of the greyhound are greater than those of the fox, and we should then equate numbers of different denominations; that is, numbers referring to different units. Hence it is necessary to express the leaps of the fox by means of those of the greyhound, or reciprocally. Now, according to the enunciation, 3 leaps of the greyhound are equivalent to 7 leaps of the fox, then 1 leap of the greyhound is equivalent to $\frac{7}{3}$ leaps of the fox, and consequently x leaps of the greyhound are equivalent to $\frac{7x}{3}$ of the fox.

Hence, we have the equation

$$\frac{7x}{3}=60+\frac{3}{2}x;$$

making the denominators disappear

$$14x=360+9x,$$

whence $\qquad 5x=360 \quad \text{and} \quad x=72.$

Therefore the greyhound will make 72 leaps to overtake the fox, and during this time the fox will make

$$72\times\frac{3}{2} \quad \text{or} \quad 108$$

Verification.

The 72 leaps of the greyhound are equivalent to

$$\frac{72\times7}{3}=168$$

leaps of the fox. And

$$60+108=168,$$

the leaps which the fox made from the beginning.

13. A father leaves his property, amounting to $2520, to four sons, A, B, C, and D. C is to have $360, B as much as C and D together, and A twice as much as B less $1000: how much does A, B, and D receive?

Ans. A $760, B $880, D $520.

14. An estate of $7500 is to be divided between a widow, two sons, and three daughters, so that each son shall receive twice as much as each daughter, and the widow herself $500 more than all the children: what was her share, and what the share of each child?

Ans. { Widow's share $4000.
Each son's $1000.
Each daughter's $500.

15. A company of 180 persons consists of men, women, and children. The men are 8 more in number than the women, and the children 20 more than the men and women together: how many of each sort in the company?

Ans. 44 men, 36 women, 100 children.

16. A father divides $2000 among five sons, so that each elder should receive $40 more than his next younger brother: what is the share of the youngest? *Ans.* $320.

17. A purse of $2850 is to be divided among three persons, A, B, and C. A's share is to be to B's as 6 to 11,

and C is to have $300 more than A and B together: what is each one's share?

Ans. A's $450, B's $825, C's $1575

18. Two pedestrians start from the same point; the first steps twice as far as the second, but the second makes 5 steps while the first makes but one. At the end of a certain time they are 300 feet apart. Now, allowing each of the longer paces to be 3 feet, how far will each have travelled?

Ans. 1st, 200 *feet*; 2nd, 500.

19. Two carpenters, 24 journeymen, and 8 apprentices, received at the end of a certain time $144. The carpenters received $1 per day, each journeyman half a dollar, and each apprentice 25 cents: how many days were they employed? *Ans.* 9 *days.*

20. A capitalist receives a yearly income of $2940: four-fifths of his money bears an interest of 4 per cent, and the remainder at 5 per cent: how much has he at interest?

Ans. 70000.

21. A cistern containing 60 gallons of water has three unequal cocks for discharging it; the largest will empty it in one hour, the second in two hours, and the third in three: in what time will the cistern be emptied if they all run together? *Ans.* $32\frac{8}{11}$ *min.*

22. In a certain orchard $\frac{1}{2}$ are apple trees, $\frac{1}{4}$ peach trees, $\frac{1}{6}$ plum trees, 120 cherry trees, and 80 pear trees: how many trees in the orchard? *Ans.* 2400.

23. A farmer being asked how many sheep he had, answered that he had them in five fields; in the 1st he had $\frac{1}{4}$, in the 2nd $\frac{1}{6}$, in the 3rd $\frac{1}{8}$, and in the 4th $\frac{1}{12}$, and in the 5th 450: how many had he? *Ans.* 1200.

24. My horse and saddle together are worth $132, and the horse is worth ten times as much as the saddle: what is the value of the horse? *Ans.* 120.

25. The rent of an estate is this year 8 per cent greater than it was last. This year it is \$1890 : what was it last year? *Ans.* \$1750.

26. What number is that from which, if 5 be subtracted, $\frac{2}{3}$ of the remainder will be 40 ? *Ans.* 65.

27. A post is $\frac{1}{4}$ in the mud, $\frac{1}{3}$ in the water, and 10 feet above the water: what is the whole length of the post? *Ans.* 24 *feet.*

28. After paying $\frac{1}{4}$ and $\frac{1}{5}$ of my money, I had 66 guineas left in my purse: how many guineas were in it at first? *Ans.* 120.

29. A person was desirous of giving 3 pence apiece to some beggars, but found he had not money enough in his pocket by 8 pence: he therefore gave them each 2 pence and had 3 pence remaining: required the number of beggars. *Ans.* 11.

30. A person in play lost $\frac{1}{4}$ of his money, and then won 3 shillings; after which he lost $\frac{1}{3}$ of what he then had; and this done, found that he had but 12 shillings remaining: what had he at first? *Ans.* 20*s.*

31. Two persons, A and B, lay out equal sums of money in trade; A gains \$126, and B loses \$87, and A's money is now double of B's: what did each lay out? *Ans.* \$300.

32. A person goes to a tavern with a certain sum of money in his pocket, where he spends 2 shillings; he then borrows as much money as he had left, and going to another tavern, he there spends 2 shillings also; then borrowing again as much money as was left, he went to a third tavern, where likewise he spent two shillings and borrowed as much as he had left; and again spending 2 shillings at a fourth tavern, he then had nothing remaining. What had he at first? *Ans.* 3*s.* 9*d.*

Of Equations of the First Degree involving two or more unknown quantities.

72. Although several of the questions hitherto resolved contained in their enunciation more than one unknown quantity, we have resolved them all by employing but one symbol. The reason of this is, that we have been able, from the conditions of the enunciation, to express easily the other unknown quantities by means of this symbol; but we are unable to do this in all problems containing more than one unknown quantity.

To ascertain how problems of this kind are resolved, let us take some of those which have been resolved by means of one unknown quantity.

1. Given the sum of two numbers equal to 36 and their difference equal to 12, to find the numbers.

Let $x=$ the greater, and $y=$ the less number.

Then, by the 1st condition . . . . . $x+y=36$,

and by the 2nd, ... $x-y=12$.

By adding (Art. **65**, Ax. 1), ... $2x=48$.

By subtracting (Art. **65**, Ax. 2), . . . $2y=24$.

Each of these equations contains but one unknown quantity

From the first we obtain . . . . $x=\frac{48}{2}=24$.

And from the second . . . . $y=\frac{24}{2}=12$

Verification.

$x+y=36$ gives $24+12=36$

$x-y=12$ „ $24-12=12$

General Solution.

Let $x=$ the greater, and y the less number.

Then by the conditions . . . . $x+y=a$,

and . . . . $x-y=b$.

By adding, (Art. **65**, Ax. 1), . . . $2x=a+b$.

By subtracting, (Art. **65**, Ax. 2), . . $2y=a-b$.

Each of these equations contains but one unknown quantity

From the first we obtain . . . . $x=\frac{a+b}{2}$.

And from the second . . . . $y=\frac{a-b}{2}$.

Verification.

$$\frac{a+b}{2}+\frac{a-b}{2}=\frac{2a}{2}=a; \text{ and } \frac{a+b}{2}-\frac{a-b}{2}=\frac{2b}{2}=b.$$

For a second example, let us also take a problem that has oeen already solved.

2. A person engaged a workman for 48 days. For each day that he labored he was to receive 24 cents, and for each day that he was idle he was to pay 12 cents for his board. At the end of the 48 days the account was settled, when the laborer received 504 cents. Required the number of working days, and the number of days he was idle.

Let $x=$ the number of working days,

$y=$ the number of idle days.

Then, $24x=$ what he earned,

and $12y=$ what he paid for his board.

Then, by the conditions of the question, we have

$$x+y\quad=48,$$

and

$$24x-12y=504.$$

This is the statement of the question

It has already been shown (Art. **65**, Ax. 3), that the two members of an equation can be multiplied by the same number, without destroying the equality. Let, then, the first equation be multiplied by 24, the coefficient of x in the second: we shall then have

$$24x+24y=1152,$$
$$24x-12y=504.$$

And by subtracting, $$36y=648,$$

and $$y=\frac{648}{36}=18.$$

Substituting this value of y in the equation

$24x-12y=504$, we have $24x-216=504$,

which gives

$24x=504+216=720$, and $x=\frac{720}{24}=30$.

Verification.

$x+y=48$ gives $30+18=48$,

$24x-12y=504$ gives $24\times30-12\times18=504$.

Elimination.

73. The method which has just been explained of combining two equations, involving two unknown quantities, and deducing therefrom a single equation involving but one, is called *elimination.*

QUEST.—**73**. What is elimination? How many methods of elimination are there? Give the rule for elimination by addition and subtraction? What is the first step? What the second? What the third?

There are three principal methods of elimination:

1st. By addition and subtraction

2d. By substitution.

3d. By comparison.

We will consider these methods separately.

Elimination by Addition and Subtraction.

1. Take the two equations

$$3x-2y=7$$
$$8x+2y=48.$$

If we add these two equations, member to member, we obtain

$$11x=55.$$

which gives, by dividing by 11

$$x=5:$$

and substituting this value in either of the given equations, we find

$$y=4.$$

2. Again, take the equations

$$8x+2y=48$$
$$3x+2y=23.$$

If we subtract the 2nd equation from the first, we obtain

$$5x=25,$$

which gives, by dividing by 5

$$x=5:$$

and by substituting this value, we find

$$y=4.$$

3. Take the two equations

$$5x+7y=43.$$
$$11x+9y=69.$$

If, in these equations, one of the unknown quantities was affected with the same coefficient, we might, by a simple subtraction, form a new equation which would contain but one unknown quantity.

Now, if both members of the first equation be multiplied by 9, the coefficient of y in the second, and the two members of the second by 7, the coefficient of y in the first, we will obtain

$$45x+63y=387,$$
$$77x+63y=483.$$

Subtracting, then, the first of these equations from the second, there results

$$32x=96, \quad \text{whence} \quad x=3.$$

Again, if we multiply both members of the first equation by 11, the coefficient of x in the second, and both members of the second by 5, the coefficient of x in the first, we will form the two equations

$$55x+77y=473,$$
$$55x+45y=345.$$

Subtracting, then, the second of these two equations from the first, there results

$$32y=128, \quad \text{whence} \quad y=4.$$

Therefore $x=3$ and $y=4$, are the values of x and y.

Verification.

$$5x+7y=43 \quad \text{gives} \quad 5\times3+7\times4=15+28=43;$$
$$11x+95=69 \quad \text{,,} \quad 11\times3+9\times4=33+36=69.$$

The method of elimination just explained, is called the *method by addition and subtraction.*

To eliminate by this method we have the following

RULE.

I. *See which of the unknown quantities you will eliminate.*

II. *Make the coefficient of this unknown quantity the same in both equations, either by multiplication or division.*

III. *If the signs of the like terms are the same in both equations, subtract one equation from the other; but if the signs are unlike, add them.*

EXAMPLES.

4. Find the values of x and y in the equations

$$3x-y=3,$$
$$y+2x=7.$$

Ans. $x=2,\ y=3.$

5. Find the values of x and y in the equations

$$4x-7y=-22,$$
$$5x+2y=37.$$

Ans. $x=5,\ y=6$

6. Find the values of x and y in the equations

$$2x+6y=42,$$
$$8x-6y=\ 3.$$

Ans. $x=4\frac{1}{2},\ y=5\frac{1}{2}.$

7. Find the values of x and y in the equations

$$8x-9y=1.$$
$$6x-3y=4x.$$

Ans. $x=\frac{1}{2},\ y=\frac{1}{3}.$

8. Find the values of x and y in the equations

$$14x-15y=12,$$
$$7x+\ 8y=37.$$

Ans. $x=3$, $y=2$.

9. Find the values of x and y in the equations

$$\frac{1}{2}x+\frac{1}{3}y=6,$$
$$\frac{1}{3}x+\frac{1}{2}y=6\tfrac{1}{2}.$$

Ans. $x=6$, $y=9$.

10. Find the values of x and y in the eqr.ations

$$\frac{1}{7}x+\frac{1}{8}y=4,$$
$$x-y=-2.$$

Ans. $x=14$, $y=16$.

11. Says A to B, you give me $40 of your money, and I shall then have 5 times as much as you will have left. Now they both had $120: how much had each?

Ans. Each had $60.

12. A Father says to his son, "twenty years ago, my age was four times yours; now, it is just double:" what wer their ages?

Ans. { Father's 60 years
Son's 30 years.

13. A Father divides his property between his two sons. At the end of the first year the elder had spent one quarter of his, and the younger had made $1000, and their property was then equal. After this the elder spends $500 and the younger makes $2000, when it appears the younger has just double the elder: what had each from the father?

Ans. { Elder $4000
Younger $2000

14. If John give Charles 15 apples, they will have the same number; but if Charles give 15 to John, John will have 15 times as many wanting 10 as Charles will have left. How many had each? *Ans.* { John 50. Charles 20.

15. Two clerks, A and B, have salaries which are together equal to \$900. A spends $\frac{1}{10}$ per year of what he receives, and B adds as much to his as A spends. At the end of the year they have equal sums: what was the salary of each? *Ans.* { A's=500. B's=400.

Elimination by Substitution.

74. Let us again take the equations

$$5x+7y=43,$$
$$11x+9y=69.$$

Find the value of x in the first equation, which gives

$$x=\frac{43-7y}{5}.$$

Substitute this value of x in the second equation, and we have

$$11\times\frac{43-7y}{5}+9y=69,$$

or, $$473-77y+45y=345,$$

or, $$-32y=-128.$$

Hence, $$y=4,$$

and, $$x=\frac{43-28}{5}=3.$$

This method is called the method by *substitution:* we have for it the following

RULE.

Find the value of one of the unknown quantities in either of the equations, and substitute this value for the same unknown quantity in the other equation: there will thus arise a new equation with but one unknown quantity.

REMARK.—This method of elimination is used to great advantage when the coefficient of either of the unknown quantities is unity.

EXAMPLES.

1. Find, by the last method, the values of x and y in the equations

$$3x-y=1 \quad \text{and} \quad 3y-2x=4$$

Ans. $x=1,\ y=2$.

2. Find the values of x and y in the equations

$$5y-4x=-22 \quad \text{and} \quad 3y+4x=38.$$

Ans. $x=8,\ y=2$.

3. Find the values of x and y in the equations

$$x+8y=18 \quad \text{and} \quad y-3x=-29.$$

Ans. $x=10,\ y=1$.

4. Find the values of x and y in the equations

$$5x-y=13 \quad \text{and} \quad 8x+\frac{2}{9}y=29.$$

Ans. $x=3\frac{1}{2},\ y=4\frac{1}{2}$.

QUEST.—**74.** Give the rule for elimination by substitution? When is it desirable to use this method?

5. Find the values of x and y from the equations

$$10x-\frac{y}{5}=69 \quad \text{and} \quad 10y-\frac{x}{7}=49.$$

Ans. $x=7,\ y=5$

6. Find the values of x and y from the equations

$$x+\frac{1}{2}x-\frac{y}{5}=10 \quad \text{and} \quad \frac{x}{8}+\frac{y}{10}=2.$$

Ans. $x=8,\ y=10.$

7. Find the values of x and y in the equations

$$\frac{y}{7}-\frac{x}{3}+5=2, \quad x+\frac{y}{5}=17\tfrac{4}{5}.$$

Ans. $x=15,\ y=14.$

8 Find the values of x and y in the equations

$$\frac{y}{2}+\frac{x}{3}+3=6\tfrac{1}{6}, \quad \text{and} \quad \frac{y}{4}-\frac{x}{7}=\frac{1}{2}.$$

Ans. $x=3\tfrac{1}{2},\ y=4.$

9. Find the values of x and y from the equations

$$\frac{y}{8}-\frac{x}{4}+6=5 \quad \text{and} \quad \frac{x}{12}-\frac{y}{16}=0.$$

Ans. $x=12,\ y=16.$

10. Find the values of x and y from the equations

$$\frac{y}{7}-\frac{3x}{2}-1=-9 \quad \text{and} \quad 5x-\frac{7y}{49}=29.$$

Ans. $x=6,\ y=7.$

11. Two misers A and B sit down to count over their money. They both have \$20000, and B has three times as much as A: how much has each?

Ans. $\begin{cases} \text{A} \ldots \ \$5000. \\ \text{B} \ldots \ \$15000 \end{cases}$

12. A person has two purses. If he puts \$7 into the first purse, it is worth three times as much as the second: but if he puts \$7 into the second it becomes worth five times as much as the first: what is the value of each purse?

Ans. 1st, \$2: 2nd, \$3.

13. Two numbers have the following properties: if the first be multiplied by 6 the product will be equal to the second multiplied by 5; and one subtracted from the first leaves the same remainder as 2 subtracted from the second: what are the numbers? *Ans.* 5 and 6.

14. Find two numbers with the following properties: the first increased by 2 to be $3\frac{1}{4}$ times greater than the second: and the second increased by 4 to be half the first: what are the numbers? *Ans.* 24 and 8.

15. A father says to his son, "twelve years ago I was twice as old as you are now: four times your age, at that time, plus twelve years, will express my age twelve years hence:" what were their ages? *Ans.* { Father 72 years. Son 30 " }

Elimination by Comparison.

75. Take the same equations

$$5x+7y=43,$$

$$11x+9y=69.$$

Finding the value of x in the first equation, we have

$$x=\frac{43-7y}{5}:$$

and finding the value of x in the second, we obtain

$$x=\frac{69-9y}{11}.$$

Let these two values of x be placed equal to each other, and we have

$$\frac{43-7y}{5}=\frac{69-9y}{11}.$$

Or, $$473-77y=345-45y;$$

Or, $$-32y=-128.$$

Hence, $$y=4.$$

And, $$x=\frac{69-36}{11}=3.$$

This method of elimination is called the method by comparison, for which we have the following

RULE.

I. *Find the value of the same unknown quantity in each equation.*

II. *Place these values equal to each other; and a new equation will arise with but one unknown quantity.*

EXAMPLES.

1. Find, by the last rule, the values of x and y in the equations

$$3x+\frac{y}{5}+6=42 \quad \text{and} \quad y-\frac{x}{22}=14\tfrac{1}{2}.$$

Ans. $x=11,\ y=15$

QUEST.—75. Give the rule for elimination by comparison? What is the first step? What the second?

2. Find the values of x and y in the equations

$$\frac{y}{4}-\frac{x}{7}+5=6 \quad \text{and} \quad \frac{y}{5}+4=\frac{x}{14}+6.$$

Ans. $x=28$, $y=20$.

3. Find the values of x and y in the equations

$$\frac{y}{10}-\frac{x}{4}+\frac{22}{8}=1 \quad \text{and} \quad 3y-x=6.$$

Ans. $x=9$, $y=5$.

4. Find the values of x and y in the equations

$$y-3=\frac{1}{2}x+5 \quad \text{and} \quad \frac{x+y}{2}=y-3\tfrac{1}{2}.$$

Ans. $x=2$, $y=9$.

5. Find the values of x and y in the equations

$$\frac{y-x}{3}+\frac{x}{2}=y-2 \quad \text{and} \quad \frac{x}{8}+\frac{y}{7}=x-13.$$

Ans. $x=16$, $y=7$.

6. Find the values of x and y from the equations

$$\frac{y+x}{2}+\frac{y-x}{2}=x-\frac{2y}{3}, \quad \text{and} \quad x+y=16.$$

Ans. $x=10$, $y=6$

7. Find the values of x and y in the equations

$$\frac{2x-3y}{5}=x-2\tfrac{2}{5}, \quad x-\frac{y-1}{2}=0.$$

Ans. $x=1$, $y=3$

8. Find the values of x and y in the equations

$$2y+3x=y+43, \quad y-\frac{x-4}{3}=y-\frac{x}{5}.$$

Ans. $x=10$, $y=13$.

9. Find the values of x and y in the equations

$$4y-\frac{x-y}{2}=x+18, \text{ and } 27-y=x+y+4.$$

Ans. $x=9$, $y=7$

10. Find the values of x and y in the equations

$$1-\frac{y-x}{6}+4=y-16\tfrac{2}{3}, \quad \frac{y}{5}-2=\frac{x}{5}.$$

Ans. $x=10$, $y=20$.

76. Having explained the principal methods of elimination, we shall add a few examples which may be solved by either; and often indeed, it may be advantageous to use them all even in the same question.

GENERAL EXAMPLES.

1. Given $2x+3y=16$, and $3x-2y=11$ to find the values of x and y. *Ans.* $x=5$, $y=2$.

2. Given $\frac{2x}{5}+\frac{3y}{4}=\frac{9}{20}$ and $\frac{3x}{4}+\frac{2y}{5}=\frac{61}{120}$ to find the values of x and y. *Ans.* $x=\frac{1}{2}$, $y=\frac{1}{3}$.

3. Given $\frac{x}{7}+7y=99$, and $\frac{y}{7}+7x=51$, to find the values of x and y. *Ans.* $x=7$, $y=14$.

4. Given

$$\frac{x}{2}-12=\frac{y}{4}+8, \text{ and } \frac{x+y}{5}+\frac{x}{3}-8=\frac{2y-x}{4}+27,$$

to find the values of x and y. *Ans.* $x=60$, $y=40$.

QUESTIONS.

1. What fraction is that, to the numerator of which, if 1 be added, its value will be $\frac{1}{3}$, but if one be added to its denominator, its value will be $\frac{1}{4}$?

Let the fraction be represented by $\frac{x}{y}$.

Then, by the question

$$\frac{x+1}{y}=\frac{1}{3} \quad \text{and} \quad \frac{x}{y+1}=\frac{1}{4}.$$

Whence $\qquad 3x+3=y \quad \text{and} \quad 4x=y+1.$

Therefore, by subtracting,

$$x-3=1 \quad \text{or} \quad x=4.$$

Hence, $\qquad 12+3=y;$

therefore, $\qquad y=15.$

2. A market-woman bought a certain number of eggs at 2 for a penny, and as many others, at 3 for a penny; and having sold them again altogether, at the rate of 5 for $2d$, found that she had lost $4d$: how many eggs had she?

Let $\qquad 2x=$ the whole number of eggs.

Then $\qquad x=$ the number of eggs of each sort.

Then will $\qquad \frac{1}{2}x=$ the cost of the first sort,

and $\qquad \frac{1}{3}x=$ the cost of the second sort.

But $5 : 2x :: 2 : \frac{4x}{5}$ the amount for which the eggs were sold.

Hence, by the question

$$\frac{1}{2}x+\frac{1}{3}x-\frac{4x}{5}=4.$$

Therefore, $\quad 15x+10x-24x=120$

or $\quad x=120;$

the number of eggs of each sort.

3. A person possessed a capital of 30,000 dollars, for which he drew a certain interest; but he owed the sum of 20,000 dollars, for which he paid a certain interest. The interest that he received exceeded that which he paid by 800 dollars. Another person possessed 35,000 dollars, for which he received interest at the second of the above rates; but he owed 24,000 dollars, for which he paid interest at the first of the above rates. The interest that he received exceeded that which he paid by 310 dollars. Required the two rates of interest.

Let x and y denote the two rates of interest; that is, the interest of $100 for the given time.

To obtain the interest of $30,000 at the first rate, denoted by x, we form the proportion

$$100 : x :: 30{,}000 : \frac{30{,}000x}{100} \quad \text{or} \quad 300x.$$

And for the interest $20,000, the rate being y,

$$100 : y :: 20{,}000 : \frac{20{,}000y}{100} \quad \text{or} \quad 200y.$$

But from the enunciation, the difference between these two interests is equal to 800 dollars.

We have, then, for the first equation of the problem

$$300x-200y=800$$

By writing algebraically the second condition of the problem, we obtain the other equation,

$$350y-240x=310.$$

Both members of the first equation being divisible by 100, and those of the second by 10, we may put the following, in place of them:

$$3x-2y=8, \qquad 35y-24x=31.$$

To eliminate x, multiply the first equation by 8, and then add it to the second; there results

$$19y=95, \quad \text{whence} \quad y=5.$$

Substituting for y, in the first equation, its value, this equation becomes

$$3x-10=8, \quad \text{whence} \quad x=6.$$

Therefore, the first rate is 6 per cent, and the second 5.

Verification.

\$30,000,	placed at 6 per cent, gives	$300\times6=\$1800.$
\$20,000,	„ 5 „ „	$200\times5=\$1000.$

And we have $1800-1000=800.$

The second condition can be verified in the same manner.

4. What two numbers are those, whose difference is 7, and sum 33? *Ans.* 13 and 20.

5. To divide the number 75 into two such parts, that three times the greater may exceed seven times the less by 15. *Ans.* 54 and 21.

6. In a mixture of wine and cider, $\frac{1}{2}$ of the whole plus 25 gallons was wine, and $\frac{1}{3}$ part minus 5 gallons was cider: how many gallons were there of each?

Ans. 85 of wine, and 35 of cider

7. A bill of £120 was paid in guineas and moidores, and the number of pieces of both sorts that were used was just 100. If the guinea be estimated at 21*s*, and the moidore at 27*s*, how many were there of each? *Ans.* 50 of each.

8. Two travellers set out at the same time from London and York, whose distance apart is 150 miles. One of them goes 8 miles a day, and the other 7: in what time will they meet? *Ans.* In 10 days.

9. At a certain election, 375 persons voted for two candidates, and the candidate chosen had a majority of 91: how many voted for each?

Ans. 233 for one, and 142 for the other.

10. A person has two horses, and a saddle worth £50. Now, if the saddle be put on the back of the first horse, it will make his value double that of the second; but if it be put on the back of the second, it will make his value triple that of the first. What is the value of each horse?

Ans. One £30, and the other £40.

11. The hour and minute hands of a clock are exactly together at 12 o'clock: when are they next together?

Ans. 1*hr.* $5\frac{5}{11}$*min.*

12. A man and his wife usually drank out a cask of beer in 12 days; but when the man was from home, it lasted the woman 30 days: how many days would the man alone be in drinking it? *Ans.* 20 days.

13. If 32 pounds of sea-water contain 1 pound of salt, how much fresh water must be added to these 32 pounds, in order that the quantity of salt contained in 32 pounds of the new mixture shall be reduced to 2 ounces, or $\frac{1}{8}$ of a pound? *Ans.* 224*lb.*

14. A person who possessed 100,000 dollars, placed the greater part of it out at 5 per cent interest, and the other

at 4 per cent. The interest which he received for the whole amounted to 4640 dollars. Required the two parts.

Ans. 64,000 and 36,000.

15. At the close of an election, the successful candidate had a majority of 1500 votes. Had a fourth of the votes of the unsuccessful candidate been also given to him, he would have received three times as many as his competitor wanting three thousand fivehundred : how many votes did each receive ?

Ans. { 1st, 6500
2d, 5000.

16. A gentlemen bought a gold and a silver watch, and a chain worth $25. When he put the chain on the gold watch, it was worth three and a half times more than the silver watch ; but when he put the chain on the silver watch, it was worth one half the gold watch and 15 dollars over : what was the value of each watch ?

Ans. { Gold watch $80.
Silver „ $30.

17. There is a certain number expressed by two figures, which figures are called digits. The sum of the digits is 11, and if 13 be added to the first digit the sum will be three times the second : what is the number ? *Ans.* 56.

18. From a company of ladies and gentlemen 15 ladies retire ; there are then left two gentlemen to each lady. After which, 45 gentlemen depart, when there are left 5 ladies to each gentleman : how many were there of each at first ?

Ans. { 50 gentlemen.
40 ladies.

19. A person wishes to dispose of his horse by lottery If he sells the tickets at $2 each, he will lose $30 on his horse ; but if he sells them at $3 each, he will receive $30 more than his horse cost him. What is the value of the horse and number of tickets ?

Ans. { Horse . . . $150.
No. of tickets 60.

20. A person purchases a lot of wheat at \$1, and a lot of rye at 75 cents per bushel, the whole costing him \$117,50. He then sells $\frac{1}{4}$ of his wheat and $\frac{1}{5}$ of his rye at the same rate, and realizes \$27,50. How much did he buy of each?

Ans. $\begin{cases} 80bu. \text{ of wheat.} \\ 50bu. \text{ of rye.} \end{cases}$

Equations involving three or more unknown quantities.

77. Let us now consider the case of three equations involving three unknown quantities.

Take the equations

$$5x-6y+4z=15,$$

$$7x+4y-3z=19,$$

$$2x+\ y+6z=46.$$

To eliminate z by means of the first two equations, multiply the first by 3 and the second by 4; then, since the coefficients of z have contrary signs, add the two results together. This gives a new equation:

$$43x-2y=121.$$

Multiplying the second equation by 2, a factor of the coefficient of z in the third equation, and adding them together, we have

$$16x+9y=84.$$

The question is then reduced to finding the values of x and y, which will satisfy these new equations.

Now, if the first be multiplied by 9, the second by 2, and the results be added together, we find

$$419x=1257, \quad \text{whence} \quad x=3$$

We might, by means of the two equations involving x and y, determine y in the same way we have determined x; but the value of y may be determined more simply, by observing that the last of these two equations becomes, by substituting for x its value found above,

$$48+9y=84, \quad \text{whence} \quad y=\frac{84-48}{9}=4.$$

In the same manner the first of the three proposed equations becomes, by substituting the values of x and y,

$$15-24+4z=15, \quad \text{whence} \quad z=\frac{24}{4}=6.$$

Hence, to solve equations containing three or more unknown quantities, we have the following

RULE.

I. *To eliminate one of the unknown quantities, combine any one of the equations with each of the others; there will thus be obtained a series of new equations containing one less unknown quantity.*

II. *Eliminate another unknown quantity by combining one of these new equations with the others.*

III. *Continue this series of operations until a single equation containing but one unknown quantity is obtained, from which the value of this unknown quantity is easily found. Then, by going back through the series of equations which have been obtained, the values of the other unknown quantities may be successively determined.*

QUEST.—**77.** Give the general rule for solving equations involving three or more unknown quantities? What is the first step? What the second? What the third?

78. REMARK.—It often happens that each of the proposed equations does not contain all the unknown quantities. In this case, with a little address, the elimination is very quickly performed.

Take the four equations involving four unknown quantities:

(1.) $2x-3y+2z=13.$ (3.) $4y+2z=14.$

(2.) $4u-2x=30.$ (4.) $5y+3u=32.$

By inspecting these equations, we see that the elimination of z in the two equations, (1) and (3), will give an equation involving x and y; and if we eliminate u in the equations (2) and (4), we shall obtain a second equation, involving x and y. These two last unknown quantities may therefore be easily determined. In the first place, the elimination of z in (1) and (3) gives

$$7y-2x=1;$$

That of u in (2) and (4), gives

$$20y+6x=38.$$

Multiplying the first of these equations by 3, and adding,

$$41y=41;$$

Whence $y=1.$

Substituting this value in $7y-2x=1$, we find

$$x=3.$$

Substituting for x its value in equation (2), it becomes

$$4u-6=30:$$

Whence $u=9.$

And substituting for y its value in equation (3), there results $z=5.$

EXAMPLES.

1. Given $\left\{\begin{array}{r} x+y+z=29 \\ x+2y+3z=62 \\ \frac{1}{2}x+\frac{1}{3}y+\frac{1}{4}z=10 \end{array}\right\}$ to find x, y and z

Ans. $x=8$, $y=9$, $z=12$.

2. Given $\left\{\begin{array}{r} 2x+4y-3z=22 \\ 4x-2y+5z=18 \\ 6x+7y-z=63 \end{array}\right\}$ to find x, y and z.

Ans. $x=3$, $y=7$, $z=4$.

3. Given $\left\{\begin{array}{r} x+\frac{1}{2}y+\frac{1}{3}z=32 \\ \frac{1}{3}x+\frac{1}{4}y+\frac{1}{5}z=15 \\ \frac{1}{4}x+\frac{1}{5}y+\frac{1}{6}z=12 \end{array}\right\}$ to find x, y and z.

Ans. $x=12$, $y=20$, $z=30$

4. Divide the number 90 into four such parts that the first increased by 2, the second diminished by 2, the third multiplied by 2, and the fourth divided by 2, shall be equal to each other.

This question may be easily solved by introducing a new unknown quantity.

Let x, y, z, and u, be the required parts, and designate by m the several equal quantities which arise from the conditions. We shall then have

$$x+2=m, \quad y-2=m, \quad 2z=m, \quad \frac{u}{2}=m.$$

From which we find

$$x=m-2,\ y=m+2,\ z=\frac{m}{2},\ u=2m.$$

And by adding the equations,

$$x+y+z+u=m+m+\frac{m}{2}+2m=4\tfrac{1}{2}m.$$

And since, by the conditions of the question, the first member is equal to 90, we have

$$4\tfrac{1}{2}m=90, \quad \text{or} \quad \tfrac{9}{2}m=90\,;$$

hence $$m=20.$$

Having the value of m, we easily find the other values: viz.

$$x=18,\ y=22,\ z=10,\ u=40.$$

5. There are three ingots composed of different metals mixed together. A pound of the first contains 7 ounces of silver, 3 ounces of copper, and 6 of pewter. A pound of the second contains 12 ounces of silver, 3 ounces of copper, and 1 of pewter, A pound of the third contains 4 ounces of silver, 7 ounces of copper, and 5 of pewter. It is required to find how much it will take of each of the three ingots to form a fourth, which shall contain in a pound, 8 ounces of silver, $3\frac{3}{4}$ of copper, and $4\frac{1}{4}$ of pewter.

Let x, y, and z represent the number of ounces which it is necessary to take from the three ingots respectively, in order to form a pound of the required ingot. Since there are 7 ounces of silver in a pound, or 16 ounces, of the first ingot, it follows that one ounce of it contains $\frac{7}{16}$ of an ounce of silver, and consequently in a number of ounces denoted by x, there is $\dfrac{7x}{16}$ ounces of silver. In the same manner we would find that $\frac{12y}{16}$ and $\frac{4z}{16}$, express the number of ounces of silver taken from the second and third, to form the fourth ; but from the enunciation, one pound of this fourth ingot contains 8 ounces of silver. We have, then, for the first equation,

$$\frac{7x}{16}+\frac{12y}{16}+\frac{4z}{16}=8;$$

or, making the denominators disappear,

$$7x+12y+4z=128.$$

As respects the copper, we should find

$$3x+3y+7z=60,$$

and with reference to the pewter

$$6x+y+5z=68.$$

As the coefficients of y in these three equations, are the most simple, it is most convenient to eliminate this unknown quantity first.

Multiplying the second equation by 4, and subtracting the first, we have

$$5x+24z=112.$$

Multiplying the third equation by 3, and subtracting the second from the product,

$$15x+8z=144.$$

Multiplying this last equation by 3, and subtracting the preceding one from the product, we obtain

$$40x=320,$$

whence $$x=8.$$

Substitute this value for x in the equation

$$15x+8z=144;$$

it becomes $$120+8z=144,$$

whence $$z=3.$$

Lastly, the two values $x=8$, $z=3$, being substituted in the equation

$$6x+y+5z=68,$$

give $$48+y+15=68,$$

whence $$y=5.$$

Therefore, in order to form a pound of the fourth ingot, we must take 8 ounces of the first, 5 ounces of the second, and 3 of the third.

Verification.

If there be 7 ounces of silver in 16 ounces of the first ingot, in 8 ounces of it, there should be a number of ounces of silver expressed by

$$\frac{7\times 8}{16}.$$

In like manner,

$$\frac{12\times 5}{16} \text{ and } \frac{4\times 3}{16}$$

will express the quantity of silver contained in 5 ounces of the second ingot, and 3 ounces of the third.

Now, we have

$$\frac{7\times 8}{16}+\frac{12\times 5}{16}+\frac{4\times 3}{16}=\frac{128}{16}=8;$$

therefore, a pound of the fourth ingot contains 8 ounces of silver, as required by the enunciation. The same conditions may be verified relative to the copper and pewter

6 A's age is double B's, and B's is triple of C's, and the sum of all their ages is 140. What is the age of each ?

Ans. A's=84, B's=42, and C's=14.

7. A person bought a chaise, horse, and harness, for £60 ; the horse came to twice the price of the harness, and the chaise to twice the price of the horse and harness What did he give for each ?

Ans. £13 6*s*. 8*d*. for the horse.
£ 6 13*s*. 4*d*. for the harness
£40 for the chaise.

8. To divide the number 36 into three such parts that $\frac{1}{2}$ of the first, $\frac{1}{3}$ of the second, and $\frac{1}{4}$ of the third, may be all equal to each other. *Ans.* 8, 12, and 16.

9. If A and B together can do a piece of work in 8 days, A and C together in 9 days, and B and C in ten days ; how many days would it take each to perform the same work alone ? *Ans.* A $14\frac{34}{49}$, B $17\frac{23}{41}$, C $23\frac{7}{31}$.

10. Three persons, A, B, and C, begin to play together, having among them all $600. At the end of the first game A has won one-half of B's money, which, added to his own, makes double the amount B had at first. In the second game, A loses and B wins just as much as C had at the beginning, when A leaves off with exactly what he had at first. How much had each at the beginning ?

Ans. A $300, B $200, C $100.

11. Three persons, A, B, and C, together possess $3640. If B gives A $400 of his money, then A will have $320 more than B ; but if B takes $140 of C's money, then B and C will have equal sums. How much has each ?

Ans. A $800, B $1280, C $1560.

12. Three persons have a bill to pay, which neither alone is able to discharge. A says to B, " Give me the 4th of your money, and then I can pay the bill." B says to C, " Give me the 8th of yours, and I can pay it. But

C says to A, "You must give me the half of yours before I can pay it, as I have but $8." What was the amount of their bill, and how much money had A and B?

Ans. { Amount of the bill, $13.
A had $10, and B $12.

13. A person possessed a certain capital, which he placed out at a certain interest. Another person, who possessed 10000 dollars more than the first, and who put out his capital 1 per cent. more advantageously, had an income greater by 800 dollars. A third person, who possessed 15000 dollars more than the first, putting out his capital 2 per cent. more advantageously, had an income greater by 1500 dollars. Required the capitals of the three persons, and the rates of interest.

Ans.				
	Sums at interest,	$30000,	40000,	45000.
	Rates of interest,	4	5	6 pr. ct.

14. A widow receives an estate of $15000 from her deceased husband, with directions to divide it among two sons and three daughters, so that each son may receive twice as much as each daughter, and she herself to receive $1000 more than all the children together. What was her share, and what the share of each child?

Ans.		
	The widow's share,	$8000.
	Each son's,	2000.
	Each daughter's,	1000.

15. A certain sum of money is to be divided between three persons, A, B, and C. A is to receive $3000 less than half of it, B $1000 less than one third part, and C to receive $800 more than the fourth part of the whole. What is the sum to be divided, and what does each receive?

Ans.		
	Sum,	$38400.
	A receives	16200.
	B "	11800.
	C "	10400.

CHAPTER IV.

Of Powers.

79. If a quantity be multiplied several times by itself the product is called the *power* of the quantity. Thus,

$a=a$	is the root, or first power of a.
$a\times a=a^2$	is the square, or second power of a.
$a\times a\times a=a^3$	is the cube, or third power of a.
$a\times a\times a\times a=a^4$	is the fourth power of a.
$a\times a\times a\times a\times a=a^5$	is the fifth power of a.

In every power there are three things to be considered :

1st. The quantity which is multiplied by itself, and which is called the *root* or the first power.

2nd. The small figure which is placed at the right, and a little above the letter. This figure is called the *exponent* of the power, and shows how many times the letter enters as a factor.

3rd. The power itself, which is the final product, or result of the multiplications.

QUEST.—**79.** If a quantity be continually multiplied by itself, what is the product called? How many things are to be considered in every power? What are they?

For example, if we suppose $a=3$, we have

$$a=3 \quad \text{the root, or 1st power of 3.}$$
$$a^2=3^2=3\times3=9 \quad \text{the second power of 3.}$$
$$a^3=3^3=3\times3\times3=27 \quad \text{the third power of 3.}$$
$$a^4=3^4=3\times3\times3\times3=81 \quad \text{the fourth power of 3.}$$
$$a^5=3^5=3\times3\times3\times3\times3=243 \quad \text{the fifth power of 3.}$$

In these expressions, 3 is the root, 1, 2, 3, 4 and 5 are the exponents, and 3, 9, 27, 81 and 243 are the powers.

To raise monomials to any power.

80. Let it be required to raise the monomial $2a^3b^2$ to the fourth power. We have

$$(2a^3b^2)^4=2a^3b^2\times2a^3b^2\times2a^3b^2\times2a^3b^2,$$

which merely expresses that the fourth power is equal to the product which arises from writing the quantity four times as a factor. By the rules for multiplication, this pro duct becomes

$$(2a^3b^2)^4=2^4a^{3+3+3+3}b^{2+2+2+2}=2^4a^{12}b^8\,;$$

from which we see,

1st. That the coefficient 2 must be raised to the 4th power; and,

2nd. That the exponent of each letter must be multiplied by 4, the exponent of the power.

As the same reasoning would apply to every example, we have, for the raising of monomials to any power, the following

RULE

I. *Raise the coefficient to the required power.*

II. *Multiply the exponent of each letter by the exponent of the power.*

EXAMPLES.

1. What is the square of $3a^2y^3$? *Ans.* $9a^4y^6$
2. What is the cube of $6a^5y^2x$? *Ans.* $216a^{15}y^6x^3$
3. What is the fourth power of $2a^3y^3b^5$? *Ans.* $16a^{12}y^{12}b^{20}$
4. What is the square of $a^2b^5y^3$? *Ans.* $a^4b^{10}y^6$.
5. What is the seventh power of a^2bcd^3? *Ans.* $a^{14}b^7c^7d^{21}$.
6. What is the sixth power of $a^2b^3c^2d$? *Ans.* $a^{12}b^{18}c^{12}d^6$.
7. What is the square and cube of $-2a^2b^2$?

Square.	*Cube.*
$-2a^2b^2$	$-2a^2b^2$
$-2a^2b^2$	$-2a^2b^2$
$+4a^4b^4$	$+4a^4b^4$
	$-2a^2b^2$
	$-8a^6b^6$.

By observing the way in which the powers are formed, we may conclude,

1st. ***When the root is positive, all the powers will be positive.***

2nd. ***When the root is negative, all the even powers will be positive and all the odd powers negative.***

QUEST.—**80**. What is a monomial? Give the rule for raising a monomial to any power. When the root is positive, how will the powers be? When the root is negative, how will the powers be?

8. What is the square of $-2a^4b^5$? *Ans.* $4a^8b^{10}$.

9. What is the cube of $-5a^5y^2c$? *Ans.* $-125a^{15}y^6c^3$.

10. What is the eighth power of $-a^3xy^2$?

Ans. $+a^{24}x^8y^{16}$

11. What is the seventh power of $-a^2yx^2$?

Ans. $-a^{14}y^7x^{14}$.

12. What is the sixth power of $2ab^6y^5$?

Ans. $64a^6b^{36}y^{30}$.

13. What is the ninth power of $-cdx^2y^3$?

Ans. $-c^9d^9x^{18}y^{27}$.

14. What is the sixth power of $-3ab^2d$?

Ans. $729a^6b^{12}d^6$.

15. What is the square of $-10a^2b^2c^3$? *Ans.* $100a^4b^4c^6$.

16, What is the cube of $-9a^6b^5d^3f^2$?

Ans. $-729a^{18}b^{15}d^9f^6$.

17. What is the fourth power of $-4a^5b^3c^4d^5$?

Ans. $256a^{20}b^{12}c^{16}d^{20}$.

18. What is the cube of $-4a^2b^2c^3d$?

Ans. $-64a^6b^6c^9d^3$

19. What is the fifth power of $2a^3b^2xy$?

Ans. $32a^{15}b^{10}x^5y^5$.

20. What is the square of $20x^4y^4c^5$? *Ans.* $400x^8y^8c^{10}$.

21. What is the fourth power of $3a^2b^2c^3$?

Ans. $81a^8b^8c^{12}$.

22. What is the fifth power of $--c^2d^3x^2y^2$?

Ans. $-c^{10}d^{15}x^{10}y^{10}$

23. What is the sixth power of $-ac^2df$?

Ans. $a^6c^{12}d^6f^6$

24. What is the fourth power of $-2a^2c^2d^3$?

Ans. $16a^8c^8d^{12}$.

To raise Polynomials to any power.

81. The power of a polynomial, like that of a monomial, is obtained by multiplying the quantity continually by itself. Thus, to find the fifth power of the binomial $a+b$, we have

$$
\begin{array}{l}
a+b \quad . \ . \ . \ . \ . \ . \ . \ . \ . \quad \text{1st power.} \\
\underline{a+b} \\
a^2+ab \\
\underline{\quad +ab+b^2} \\
a^2+2ab+b^2 \quad . \ . \ . \ . \ . \ . \ . \quad \text{2nd power.} \\
\underline{a+b} \\
a^3+2a^2b+ab^2 \\
\underline{\quad +a^2b+2ab^2+b^3} \\
a^3+3a^2b+3ab^2+b^3 \quad . \ . \ . \quad \text{3rd power.} \\
\underline{a+b} \\
a^4+3a^3b+3a^2b^2+ab^3 \\
\underline{\quad +a^3b+3a^2b^2+3ab^3+b^4} \\
a^4+4a^3b+6a^2b^2+4ab^3+b^4 \quad \text{4th power} \\
\underline{a+b} \\
a^5+4a^4b+6a^3b^2+4a^2b^3+ab^4 \\
\underline{\quad +a^4b+4a^3b^2+6a^2b^3+4ab^4+b^5} \\
a^5+5a^4b+10a^3b^2+10a^2b^3+5ab^4+b^5 \quad \textit{Ans.}
\end{array}
$$

REMARK.—**82.** It will be observed that the number of multiplications is always 1 less than the units in the expo-

---

QUEST.—81. How is the power of a polynomial obtained?

nent of the power. Thus, if the exponent is 1, no multiplication is necessary. If it is 2, we multiply once; if it is 3, twice; if 4, three times, &c. The powers of polynomials may be expressed by means of the exponent. Thus, to express that $a+b$ is to be raised to the 5th power, we write

$$(a+b)^5,$$

which expresses the fifth power of $a+b$.

2. Find the 5th power of the binomial $a-b$.

$$\begin{array}{lr}
a-b \quad \ldots\ldots\ldots & \text{1st power.} \\
\underline{a-b} & \\
a^2-ab & \\
\underline{\quad -ab+b^2} & \\
a^2-2ab+b^2 \quad \ldots\ldots & \text{2nd power.} \\
\underline{a-b} & \\
a^3-2a^2b+ab^2 & \\
\underline{\quad -a^2b+2ab^2-b^3} & \\
a^3-3a^2b+3ab^2-b^3 \quad \ldots & \text{3rd power.} \\
\underline{a-b} & \\
a^4-3a^3b+3a^2b^2-ab^3 & \\
\underline{\quad -a^3b+3a^2b^2-3ab^3+b^4} & \\
a^4-4a^3b+6a^2b^2-4ab^3+b^4 & \text{4th power.} \\
\underline{a-b} & \\
a^5-4a^4b+6a^3b^2-4a^2b^3+ab^4 & \\
\underline{\quad -a^4b+4a^3b^2-6a^2b^3+4ab^4-b^5} & \\
a^5-5a^4b+10a^3b^2-10a^2b^3+5ab^4-b^5 & \textit{Ans.}
\end{array}$$

QUEST.—**82.** How does the number of multiplications compare with the exponent of the power? If the exponent is 4, how many multiplications?

3. What is the square of $5a-2c+d$.

$$\begin{array}{l} 5a-2c+d \\ 5a-2c+d \\ \hline 25a^2-10ac+5ad \\ \quad -10ac+4c^2-2cd \\ \quad\quad +5ad-2cd+d^2 \\ \hline 25a^2-20ac+10ad+4c^2-4cd+d^2 \quad \textit{Ans.} \end{array}$$

4. Find the 4th power of the binomial $3a-2b$.

$$\begin{array}{ll} 3a-2b \ldots\ldots\ldots & \text{1st power.} \\ 3a-2b & \\ \hline 9a^2-6ab & \\ \quad -6ab+4b^2 & \\ \hline 9a^2-12ab+4b^2 \ldots\ldots & \text{2nd power.} \\ 3a-2b & \\ \hline 27a^3-36a^2b+12ab^2 & \\ \quad -18a^2b+24ab^2-8b^3 & \\ \hline 27a^3-54a^2b+36ab^2-8b^3 \ldots & \text{3rd power.} \\ 3a-2b & \\ \hline 81a^4-162a^3b+108a^2b^2-24ab^3 & \\ \quad -54a^3b+108a^2b^2-72ab^3+16b^4 & \\ \hline 81a^4-216a^3b+216a^2b^2-96ab^3+16b^4 & \textit{Ans.} \end{array}$$

5. What is the square of the binomial $a+1$?

Ans. a^2+2a+1.

6. What is the square of the binomial $a-1$?

Ans. a^2-2a+1.

7. What is the cube of $9a-3b$?

Ans. $729a^3-729a^2b+243ab^2-27b^3$.

8. What is the third power of $a-1$?

Ans. a^3-3a^2+3a-1.

9 What is the 4th power of $x-y$?

Ans. $x^4-4x^3y+6x^2y^2-4xy^3+y^4$.

10. What is the cube of the trinomial $x+y+z$?

Ans. $x^3+3x^2y+3x^2z+3xy^2+3xz^2+3y^2z+3yz^2+6xyz$
$+y^3+z^3$.

11. What is the cube of the trinomial $2a^2-4ab+3b^2$?

Ans. $8a^6-48a^5b+132a^4b^2-208a^3b^3+198a^2b^4-108ab^5$
$+27b^6$

To raise a Fraction to any Power.

83. The power of a fraction is obtained by multiplying the fraction by itself; that is, by multiplying the numerator by the numerator, and the denominator by the denominator

Thus, the cube of $\frac{a}{b}$, which is written

$$\left(\frac{a}{b}\right)^3=\frac{a}{b}\times\frac{a}{b}\times\frac{a}{b}=\frac{a^3}{b^3},$$

is found by cubing the numerator and denominator separately.

2. What is the square of the fraction $\frac{a-c}{b+c}$?

We have

$$\left(\frac{a-c}{b+c}\right)^2=\frac{(a-c)^2}{(b+c)^2}=\frac{a^2-2ac+c^2}{b^2+2bc+c^2} \quad \textit{Ans.}$$

3. What is the cube of $\frac{xy}{3bc}$? *Ans.* $\frac{x^3y^3}{27b^3c^3}$

QUEST.—**83** How do you find the power of a fraction?

4. What is the fourth power of $\frac{ab^2c}{2x^2y^2}$?

Ans. $\frac{a^4b^8c^4}{16x^8y^8}$

5. What is the cube of $\frac{x-y}{x+y}$?

Ans. $\frac{x^3-3x^2y+3xy^2-y^3}{x^3+3x^2y+3xy^2+y^3}$.

6 What is the fourth power of $\frac{2ax}{4ay}$? Ans. $\frac{x^4}{16y^4}$.

7. What is the fifth power of $\frac{9bcx}{18yz}$? Ans. $\frac{b^5c^5x^5}{32y^5z^5}$.

8. What is the square of $\frac{ax-y}{by-x}$?

Ans. $\frac{a^2x^2-2axy+y^2}{b^2y^2-2bxy+x^2}$.

9. What is the cube of $\frac{2a-3b}{x+2y}$?

Ans. $\frac{8a^3-36a^2b+54ab^2-27b^3}{x^3+6x^2y+12xy^2+8y^3}$.

Binomial Theorem.

84. The method which has been explained of raising a polynomial to any power, is somewhat tedious, and hence other methods, less difficult, have been anxiously sought for. The most simple which has yet been discovered, is the one invented by Sir Isaac Newton, called the *Binomial Theorem.*

QUEST.—84. What is the object of the Binomial Theorem? Who discovered this theorem?

85. In raising a quantity to any power, it is plain that there are four things to be considered :—

1st. The number of terms of the power.

2nd. The signs of the terms.

3rd. The exponents of the letters.

4th. The coefficients of the terms.

Of the Terms.

86. If we take the two examples of Article **81**, which we there wrought out in full; we have

$$(a+b)^5=a^5+5a^4b+10a^3b^2+10a^2b^3+5ab^4+b^5;$$

$$(a-b)^5=a^5-5a^4b+10a^3b^2-10a^2b^3+5ab^4-b^5.$$

By examining the several multiplications, in Art. **81**, we shall observe that the second power of a binomial contains three terms, the third power four, the fourth power five, the fifth power six, &c; and hence we may conclude—*That the number of terms in any power of a binomial, is one greater than the exponent of the power.*

Of the Signs of the Terms.

87. It is evident that when both terms of the given binomial are plus, *all the terms of the power will be plus.*

2nd. If the second term of the binomial is negative, then *all the odd terms, counted from the left, will be positive, and all the even terms negative.*

QUEST.—**85**. In raising a quantity to any power, how many things are to be considered? What are they?—**86**. How many terms are there in any power of a binomial? If the exponent is 3, how many terms? If it is 4, how many terms? If 5? &c.—**87**. If both terms of the binomial are positive, how are the terms of the power? If the second term is negative, how are the signs of the terms?

Of the Exponents.

88. The letter which occupies the first place in a binomial, is called the *leading letter.* Thus, a is the leading letter in the binomials $a+b$, $a-b$.

1st. It is evident that the exponent of the leading letter in the first term will be the same as the exponent of the power; and that this exponent will diminish by unity in each term to the right, until we reach the last term, which does not contain the leading letter.

2nd. The exponent of the second letter is 1 in the second term, and increases by unity in each term to the right until we reach the last term, in which the exponent is the same as that of the given power.

3rd. The sum of the exponents of the two letters, in any term, is equal to the exponent of the given power. This last remark will enable us to verify any result obtained by the binomial theorem.

Let us now apply these principles in the two following examples, in which the coefficients are omitted:—

$$(a+b)^6 \quad . \; . \; . \quad a^6+a^5b+a^4b^2+a^3b^3+a^2b^4+ab^5+b^6,$$

$$(a-b)^6 \quad . \; . \; . \quad a^6-a^5b+a^4b^2-a^3b^3+a^2b^4-ab^5+b^6.$$

As the pupil should be practised in writing the terms, with their proper signs, without the coefficients, we will add a few more examples.

QUEST.—**88**. Which is the leading letter of the binomial? What is the exponent of this letter in the first term? How does it change in the terms towards the right? What is the exponent of the second letter in the second term? How does it change in the terms towards the right? What is it in the last term? What is the sum of the exponents in any term equal to?

1. $(a+b)^3$ . . $a^3+a^2b+ab^2+b^3$.
2. $(a-b)^4$ . . $a^4-a^3b+a^2b^2-ab^3+b^4$.
3. $(a+b)^5$ . . $a^5+a^4b+a^3b^2+a^2b^3+ab^4+b^5$.
4. $(a-b)^7$ . . $a^7-a^6b+a^5b^2-a^4b^3+a^3b^4-a^2b^5+ab^6-b^7$

Of the Coefficients.

89. The coefficient of the first term is unity. The coefficient of the second term is the same as the exponent of the given power. The coefficient of the third term is found by multiplying the coefficient of the second term by the exponent of the leading letter, and dividing the product by 2. And finally—*If the coefficient of any term be multiplied by the exponent of the leading letter, and the product divided by the number which marks the place of that term from the left, the quotient will be the coefficient of the next term.*

Thus, to find the coefficients in the example

$$(a-b)^7 \ . \ . \ . \ a^7-a^6b+a^5b^2-a^4b^3+a^3b^4-a^2b^5+ab^6-b^7$$

we first place the exponent 7 as a coefficient of the second term. Then, to find the coefficient of the third term, we multiply 7 by 6, the exponent of a, and divide by 2. The quotient 21 is the coefficient of the third term. To find the coefficient of the fourth, we multiply 21 by 5, and divide the product by 3 : this gives 35. To find the coefficient of the fifth term, we multiply 35 by 4, and divide the product by 4 : this gives 35. The coefficient of the sixth term, found in the same way, is 21 ; that of the seventh, 7 ; and that of the eighth, 1. Collecting these coefficients, we have

$$(a-b)^7=$$
$$a^7-7a^6b+21a^5b^2-35a^4b^3+35a^3b^4-21a^2b^5+7ab^6-b^7.$$

REMARK.—We see, in examining this last result, that the coefficients of the extreme terms are each unity, and that the coefficients of terms equally distant from the extreme terms are equal. It will, therefore, be sufficient to find the coefficients of the first half of the terms, from which the others may be immediately written.

EXAMPLES.

1. Find the fourth power of $a+b$.

Ans. $a^4+4a^3b+6a^2b^2+4ab^3+b^4$.

2. Find the fourth power of $a-b$.

Ans. $a^4-4a^3b+6a^2b^2-4ab^3+b^4$

3. Find the fifth power of $a+b$.

Ans. $a^5+5a^4b+10a^3b^2+10a^2b^3+5ab^4+b^5$.

4. Find the fifth power of $a-b$.

Ans. $a^5-5a^4b+10a^3b^2-10a^2b^3+5ab^4-b^5$.

5. Find the sixth power of $a+b$.

Ans. $a^6+6a^5b+15a^4b^2+20a^3b^3+15a^2b^4+6ab^5+b^6$.

6. Find the sixth power of $a-b$.

Ans. $a^6-6a^5b+15a^4b^2-20a^3b^3+15a^2b^4-6ab^5+b^6$

7. Let it be required to raise the binomial $3a^2c-2bd$ to the fourth power.

It frequently occurs that the terms of the binomial are affected with coefficients and exponents, as in the above

---

QUEST.—**89.** What is the coefficient of the first term? What is the coefficient of the second? How do you find the coefficient of the third term? How do you find the coefficient of any term? What are the coefficients of the first and last terms? How are the coefficients of terms equally distant from the extremes?

example. In the first place, we represent each term of the binomial by a single letter. Thus, we place

$$3a^2c=x, \quad \text{and} \quad -2bd=y,$$

we then have

$$(x+y)^4=x^4+4x^3y+6x^2y^2+4xy^3+y^4.$$

But, $\quad x^2=9a^4c^2, \quad x^3=27a^6c^3, \quad x^4=81a^8c^4$;

and $\quad y^2=4b^2d^2, y^3=-8b^3d^3, \quad y^4=16b^4d^4.$

Substituting for x and y their values, we have

$$(3a^3c-2bd)^4=(3a^2c)^4+4(3a^2c)^3(-2bd)+6\,(3a^2c)^2\,(-2bd)^2$$
$$+4(3a^2c)\,(-2bd)^3+(-2bd)^4,$$

and by performing the operations indicated,

$$(3a^3c-2bd)^4=81a^8c^4-216a^6c^3bd+216a^4c^2b^2d^2-96a^2cb^3d^3$$
$$+16b^4d^4.$$

8. What is the square of $3a-6b$?

Ans. $9a^2-36ab+36b^2.$

9. What is the cube of $3x-6y$?

Ans. $27x^3-162x^2y+324xy^2-216y^3.$

10. What is the square of $x-y$?

Ans. $x^2-2xy+y^2.$

11. What is the eighth power of $m+n$?

Ans. $m^8+8m^7n+28m^6n^2+56m^5n^3+70m^4n^4+56m^3n^5.$
$+28m^2n^6+8mn^7+n^8.$

12. What is the fourth power of $a-3b$?

Ans. $a^4-12a^3b+54a^2b^2-108ab^3+81b^4.$

13. What is the fifth power of $c-2d$?

Ans. $c^5-10c^4d+40c^3d^2-80c^2d^3+80cd^4-32d^5.$

14. What is the cube of $5a-3d$?

Ans. $125a^3-225a^2d+135ad^2-27d^3$

REMARK. The powers of any polynomial may easily be found by the Binomial Theorem.

15. For example, raise $a+b+c$ to the third power.

First, put . . . . $b+c=d$

Then, $(a+b+c)^3=(a+d)^3=a^3+3a^2d+3ad^2+d^3$.

Or, by substituting for the value of d,

$$
\begin{aligned}
(a+b+c)^3=a^3+3a^2b+3ab^2+b^3 \\
3a^2c+3b^2c+6abc \\
+3ac^2+3bc^2 \\
+\ c^3.
\end{aligned}
$$

This expression is composed of *the cubes of the three terms, plus three times the square of each term by the first powers of the two others, plus six times the product of all three terms.* It is easily proved that this *law* is true for any polynomial.

To apply the preceding formula to the development of the cube of a trinomial, in which the terms are affected with coefficients and exponents, *designate each term by a single letter, then replace the letters introduced, by their values, and perform the operations indicated.*

From this rule, we find that

$$
(2a^2-4ab+3b^2)^3=8a^6-48a^5b+132a^4b^2-208a^3b^3
$$
$$
+198a^2b^4-108ab^5+27b^6.
$$

The fourth, fifth, &c, powers of any polynomial can be found in a similar manner.

16. What is the cube of $a-2b+c$?

Ans. $a^3-8b^3+c^3-6a^2b+3a^2c+12ab^2+12b^2c+3ac^2$
$-6bc^2-12abc$.

CHAPTER V.

Extraction of the Square Root of Numbers. Formation of the Square and Extraction of the Square Root of Algebraic Quantities. Calculus of Radicals of the Second Degree.

90. The *square* or second power of a number, is the product which arises from multiplying that number by itself once: for example, 49 is the square of 7, and 144 is the square of 12.

91. The *square root* of a number is that number which, being multiplied by itself once, will produce the given number. Thus, 7 is the square root of 49, and 12 the square root of 144: for, $7\times7=49$, and $12\times12=144$.

92. The square of a number, either entire or fractional, is easily found, being always obtained by multiplying this number by itself once. The extraction of the square root of a number is, however, attended with some difficulty, and requires particular explanation.

QUEST.—**90.** What is the square, or second power of a number?—91 What is the square root of a number?

The first ten numbers are.

1, 2, 3, 4, 5, 6, 7, 8, 9, 10;

and their squares,

1, 4, 9, 16. 25, 36, 49, 64, 81, 100;

and reciprocally, the numbers of the first line are the square roots of the corresponding numbers of the second. We may also remark that, *the square of a number expressed by a single figure, will contain no figure of a higher denomination than tens.*

The numbers of the last line, 1, 4, 9, 16, &c, and all other numbers which can be produced by the multiplication of a number by itself, are called *perfect squares.*

It is obvious that there are but nine perfect squares among all the numbers which can be expressed by one or two figures: the square roots of all other numbers expressed by one or two figures, will be found between two whole numbers differing from each other by unity. Thus 55, which is comprised between 49 and 64, has for its square root a number between 7 and 8. Also 91, which is comprised between 81 and 100, has for its square root a number between 9 and 10.

93. Every number may be regarded as made up of a certain number of tens and a certain number of units. Thus 64 is made up of 6 tens and 4 units, and may be expressed under the form $60+4$.

QUEST.—**92.** What will be the highest denomination of the square of a number expressed by a single figure? What are perfect squares? How many are there between 1 and 100? What are they?

Now, if we represent the tens by a and the units by b, we shall have

$$a+b=64,$$

and $$(a+b)^2=(64)^2;$$

or $$a^2+2ab+b^2=4096.$$

Which proves that the square of a number composed of tens and units, contains *the square of the tens plus twice the product of the tens by the units, plus the square of the units.*

94. If, now, we make the units 1, 2, 3, 4, &c, tens, or units of the second order, by annexing to each figure a cipher, we shall have

10, 20, 30, 40, 50, 60, 70, 80, 90, 100,

and for their squares,

100, 400, 900, 1600, 2500, 3600, 4900, 6400, 8100, 10000.

From which we see that the square of one ten is 100, the square of two tens 400 ; and generally, *that the square of tens will contain no figures of a less denomination than hundreds, nor of a higher name than thousands.*

Ex. 1.—To extract the square root of 6084.

Since this number is composed of more than two places of figures, its root will contain more than one. But since it is less than 10000, which is the square of 100, the root will contain but two figures : that is, units and tens.

$$\dot{6}0\,8\dot{4}$$

Now, the square of the tens must be found in the two

QUEST.—**93**. How may every number be regarded as made up? What is the square of a number composed of tens and units equal to?—**94**. What is the square of one ten equal to? Of 2 tens? Of 3 tens? &c.

left-hand figures, which we will separate from the other two by putting a point over the place of units, and a second over the place of hundreds. These parts, of two figures each, are called *periods.* The part 60 is comprised between the two squares 49 and 64, of which the roots are 7 and 8 : hence, 7 *is the figure of the tens sought;* and the required root is composed of 7 tens and a certain number of units.

The figure 7 being found, we write it on the right of the given number, from which we separate it by a vertical line : then we subtract its square, 49, from 60, which leaves a remainder of 11, to which we bring down the two next figures 84. The result of this operation, 1184, contains *twice the product of the tens by the units, plus the square of the units.*

```
               6̇0 8̇4 | 78
               49     |
7×2=14 8 |  118 4
         |  118 4
         -----------
                0
```

But since tens multiplied by units cannot give a product of a less name than tens, it follows that the last figure, 4, can form no part of the double product of the tens by the units: this double product is therefore found in the part 118, which we separate from the units' place, 4.

Now if we double the tens, which gives 14, and then divide 118 by 14, the quotient 8 *is the figure of the units*, or a figure greater than the units. This quotient figure can never be too small, since the part 118 will be at least equal to twice the product of the tens by the units : but it may be too large ; for the 118, besides the double product of the tens by the units, may likewise contain tens arising from the square of the units. To ascertain if the quotient 8 expresses the units, we write the 8 on the right of the 14, which gives 148, and then we multiply 148 by 8. Thus, we evidently form, 1st, the square of the units ; and, 2nd, the double product of the tens by the units. This multiplication being effected, gives for a product 1184, a number equal to the result of the first operation. Having subtracted the product, we find the remainder equal to 0: hence 78 is the root required.

Indeed, in the operations, we have merely subtracted from the given number 6084, 1st, the square of 7 tens, or 70; 2nd, twice the product of 70 by 8; and, 3d, the square of 8: that is, the three parts which enter into the composition of the square $70+8$, or 78; and since the result of the subtraction is 0, it follows that 78 is the square root of 6084.

95. Remark.—The operations in the last example have been performed on but two periods, but it is plain that the same reasoning is equally applicable to larger numbers, for by changing the order of the units, we do not change the relation in which they stand to each other.

Thus, in the number 60 84 95, the two periods 60 84 have the same relation to each other as in the number 60 84; and hence the methods used in the last example are equally applicable to larger numbers.

96. Hence, for the extraction of the square root of numbers, we have the following

RULE.

I. *Separate the given number into periods of two figures each, beginning at the right hand:—the period on the left will often contain but one figure.*

II. *Find the greatest square in the first period on the left, and place its root on the right, after the manner of a quotient*

Quest.—95. Will the reasoning in the example apply to more than two periods?

in division. Subtract the square of the root from the first period, and to the remainder bring down the second period for a dividend.

III. *Double the root already found, and place it on the left for a divisor. Seek how many times the divisor is contained in the dividend, exclusive of the right-hand figure, and place the figure in the root and also at the right of the divisor.*

IV. *Multiply the divisor, thus augmented, by the last figure of the root, and subtract the product from the dividend, and to the remainder bring down the next period for a new dividend But if any of the products should be greater than the dividend, diminish the last figure of the root.*

V. *Double the whole root already found, for a new divisor, and continue the operation as before, until all the periods are brought down.*

97. 1st Remark. If, after all the periods are brought down, there is no remainder, the proposed number is a perfect square. But if there is a remainder, you have only found the root of the greatest perfect square contained in the given number; or *the entire part of the root sought.*

For example, if it were required to extract the square root of 665, we should find 25 for the entire part of the root, and a remainder of 40, which shows that 665 is not a perfect square. But is the square of 25 the greatest perfect square contained in 665 ? that is, is 25 the entire part of the root ? To prove this, we will first show that, *the difference between the squares of two consecutive numbers, is equal to twice the less number augmented by unity.*

Quest.—**96.** Give the rule for extracting the square root of numbers What is the first step ? What the second ? What the third ? Wha the fourth ? What the fifth ?

Let . . $a=$ the less number,
and . . $a+1=$ the greater.
Then . $(a+1)^2=a^2+2a+1$,
and . . $(a)^2=a^2$.

Their difference is $=$ $2a+1$ as enunciated.

Hence, the entire part of the root cannot be augmented, unless the remainder exceeds twice the root found, plus unity.

But $25\times2+1=51>40$ the remainder: therefore, 25 is the entire part of the root.

98. 2nd Remark.—The number of figures in the root will always be equal to the number of periods into which the given number is separated.

EXAMPLES.

1. To find the square root of 7225. *Ans.* 85.
2. To find the square root of 17689. *Ans.* 133.
3. To find the square root of 994009. *Ans.* 997.
4. To find the square root of 85673536. *Ans.* 9256.
5. To find the square root of 67798756. *Ans.* 8234.
6. To find the square root of 978121. *Ans.* 989.
7. To find the square root of 956484. *Ans.* 978.
8. What is the square root of 36372961? *Ans.* 6031.
9. What is the square root of 22071204? *Ans.* 4698.
10. What is the square root of 106929? *Ans.* 327.
11. What is the square root of 12088868379025? *Ans.* 3476905.

Quest.—**98.** How many figures will you always find in the root?

99. 3rd REMARK.—If the given number has not an exact root, there will be a remainder after all the periods are brought down, in which case ciphers may be annexed, forming new periods, each of which will give one decimal place in the root.

1. What is the square root of 36729 ?

```
              3 67 29 | 191,64+
              1       |
           -----------------
        2 9 | 267
            | 261
           ------
       38 1 | 629
            | 381
           ------
      382 6 | 24800
            | 22956
           --------
     3832 4 | 184400
            | 153296
            --------
              31104 Rem.
```

In this example there are two periods of decimals, which give two places of decimals in the root.

2. What is the square root of 2268741 ?

Ans. 1506,23+.

3. What is the square root of 7596796 ?

Ans. 2756,22+.

4. What is the square root of 96 ?

Ans. 9,79795+.

5. What is the square root of 153 ?

Ans. 12,36931+.

6. What is the square root of 101 ?

Ans. 10,04987+.

QUEST.—**99**. How will you find the decimal part of the root ?

7. What is the square root of 285970396644 ?

Ans. 534762.

8. What is the square root of 41605800625 ?

Ans. 203975.

9. What is the square root of 48303584206084 ?

Ans. 6950078.

Extraction of the square root of Fractions.

100. Since the square or second power of a fraction is obtained by squaring the numerator and denominator separately, it follows that the square root of a fraction will be equal to the square root of the numerator divided by the square root of the denominator.

For example, the square root of $\frac{a^2}{b^2}$ is equal to $\frac{a}{b}$: for

$$\frac{a}{b} \times \frac{a}{b} = \frac{a^2}{b^2}.$$

1. What is the square root of $\frac{1}{4}$? *Ans.* $\frac{1}{2}$.

2. What is the square root of $\frac{9}{16}$? *Ans.* $\frac{3}{4}$.

3. What is the square root of $\frac{64}{81}$? *Ans.* $\frac{8}{9}$.

4. What is the square root of $\frac{256}{361}$? *Ans.* $\frac{16}{19}$.

5. What is the square root of $\frac{16}{64}$? *Ans.* $\frac{1}{2}$.

QUEST.—**100.** If the numerator and denominator of a fraction are perfect squares, how will you extract the square root ?

6. What is the square root of $\frac{4096}{61009}$? *Ans.* $\frac{64}{247}$

7. What is the square root of $\frac{582169}{956484}$? *Ans.* $\frac{763}{978}$

101. If neither the numerator nor the denominator is a perfect square, the root of the fraction cannot be exactly found. We can, however, easily find the approximate root. For this purpose,

Multiply both terms of the fraction by the denominator, which makes the denominator a perfect square without altering the value of the fraction. Then, extract the square root of the numerator, and divide this root by the root of the denomitor; this quotient will be the approximate root.

Thus, if it be required to extract the square root of $\frac{3}{5}$, we multiply both terms by 5, which gives $\frac{15}{25}$.

We then have

$$\sqrt{15} = 3{,}8729+ :$$

hence, $3{,}8729+ \div 5 = {,}7745+ =$ *Ans.*

2. What is the square root of $\frac{7}{4}$? *Ans.* 1,32287+.

3. What is the square root of $\frac{14}{9}$? *Ans.* 1,24721+.

4. What is the square root of $11\frac{11}{16}$?

Ans. 3,41869+

QUEST.—**101**. If the numerator and denominator of a fraction are not perfect squares, how do you extract the square root?

5. What is the square root of $7\frac{13}{36}$? *Ans.* 2,71313+.

6. What is the square root of $8\frac{15}{49}$? *Ans.* 2,88203+.

7. What is the square root of $\frac{5}{12}$? *Ans.* 0,64549+.

8. What is the square root of $10\frac{3}{10}$?

Ans. 3,20936+.

102. Finally, instead of the last method, we may, if we please,

Change the vulgar fraction into a decimal, and continue the division until the number of decimal places is double the number of places required in the root. Then, extract the root of the decimal by the last rule.

Ex. 1. Extract the square root of $\frac{11}{14}$ to within ,001. This number, reduced to decimals, is 0.785714 to within 0,000001; but the root of 0,785714 to the nearest unit, is ,886; hence 0,886 is the root of $\frac{11}{14}$ to within ,001.

2. Find the $\sqrt{2\frac{13}{15}}$ to within 0,0001.

Ans. 1,6931+

3. What is the square root of $\frac{1}{17}$? *Ans.* 0,24253+

4. What is the square root of $\frac{7}{8}$? *Ans.* 0,93541+.

5. What is the square root of $\frac{5}{3}$? *Ans.* 1,29099+.

QUEST.—**102.** By what other method may the root be found?

Extraction of the Square Root of Monomials.

103. In order to discover the process for extracting the square root, we must see how the square of the monomial is formed.

By the rule for the multiplication of monomials (Art. 35), we have

$$(5a^2b^3c)^2 = 5a^2b^3c \times 5a^2b^3c = 25a^4b^6c^2;$$

that is, in order to square a monomial, it is necessary to *square its coefficient, and double each of the exponents of the different letters.* Hence, to find the root of the square of a monomial, we have the following

RULE.

I. *Extract the square root of the coefficient.*

II. *Divide the exponent of each letter by 2.*

Thus, $\sqrt{64a^6b^4} = 8a^3b^2$ for $8a^3b^2 \times 8a^3b^2 = 64a^6b^4$.

2. Find the square root of $625a^2b^8c^6$. *Ans.* $25ab^4c^3$.
3. Find the square root of $576a^4b^6c^8$. *Ans.* $24a^2b^3c^4$.
4. Find the square root of $196x^6y^2z^4$. *Ans.* $14x^3yz^2$.
5. Find the square root of $441a^8b^6c^{10}d^{16}$. *Ans.* $21a^4b^3c^5d^8$.
6. Find the square root of $784a^{12}b^{14}c^{16}d^2$. *Ans.* $28a^6b^7c^8d$.
7. Find the square root of $81a^8b^4c^6$. *Ans.* $9a^4b^2c^3$.

QUEST.—**103.** How do you extract the square root of a monomial?

104. From the preceding rule it follows, that when a monomial is a perfect square, *its numerical coefficient is a perfect square, and all its exponents even numbers.* Thus, $25a^4b^2$ is a perfect square, but $98ab^4$ is not a perfect square, because 98 is not a perfect square, and a is affected with an uneven exponent.

In the latter case, the quantity is introduced into the calculus by affecting it with the sign $\sqrt{}$, and it is written thus:

$$\sqrt{98ab^4}.$$

Quantities of this kind are called *radical quantities*, or *irrational quantities*, or simply *radicals of the second degree.* They are also, sometimes called *Surds.*

These expressions may often be simplified, upon the principle that, *the square root of the product of two or more factors is equal to the product of the square roots of these factors*; or, in algebraic language,

$$\sqrt{abcd}\ .\ .\ .\ =\sqrt{a}\ .\ \sqrt{b}\ .\ \sqrt{c}\ .\ \sqrt{d}\ .\ .\ .$$

This being the case, the above expression, $\sqrt{98ab^4}$, can be put under the form

$$\sqrt{49b^4\times 2a}=\sqrt{49b^4}\times\sqrt{2a}.$$

Now $\sqrt{49b^4}$ may be reduced to $7b^2$; hence,

$$\sqrt{98ab^4}=7b^2\sqrt{2a}.$$

In like manner,

$$\sqrt{45a^2b^3c^2d}=\sqrt{9a^2b^2c^2\times 5bd}=3abc\sqrt{5bd},$$

$$\sqrt{864a^2b^5c^{11}}=\sqrt{144a^2b^4c^{10}\times 6bc}=12ab^2c^5\sqrt{6bc}$$

The quantity which stands without the radical sign is called the *coefficient* of the radical. Thus, in the expressions

$$7b^2\sqrt{2a}, \quad 3abc\sqrt{5bd}, \quad 12ab^2c^5\sqrt{6bc},$$

the quantities $7b^2$, $3abc$, $12ab^2c^5$, are called *coefficients of the radicals.*

Hence, to simplify a radical expression of the second degree, we have the following

RULE.

I. *Separate the expression into two parts, of which one shall contain all the factors that are perfect squares, and the other the remaining ones.*

II. *Take the roots of the perfect squares and place them before the radical sign, under which leave those factors which are not perfect squares.*

105. Remark.—To determine if a given number has any factor which is a perfect square, we examine and see if it is divisible by either of the perfect squares

$$4, \quad 9, \quad 16, \quad 25, \quad 36, \quad 49, \quad 64, \quad 81, \quad \&c;$$

and if it is not, we conclude that it does not contain a factor which is a perfect square.

Quest.—**104.** When is a monomial a perfect square? When it is not a perfect square, how is it introduced into the calculus? What are quantities of this kind called? May they be simplified? Upon what principle? What is a coefficient of a radical? Give the rule for reducing radicals.—**105.** How do you determine whether a given number has a factor which is a perfect square?

EXAMPLES.

1. Reduce $\sqrt{75a^3bc}$ to its simplest form.

Ans. $5a\sqrt{3abc}$.

2. Reduce $\sqrt{128b^5a^6d^2}$ to its simplest form.

Ans. $8b^2a^3d\sqrt{2b}$.

3. Reduce $\sqrt{32a^9b^8c}$ to its simplest form.

Ans. $4a^4b^4\sqrt{2ac}$.

4. Reduce $\sqrt{256a^2b^4c^8}$ to its simplest form.

Ans. $16ab^2c^4$.

5. Reduce $\sqrt{1024a^9b^7c^5}$ to its simplest form.

Ans. $32a^4b^3c^2\sqrt{abc}$.

6. Reduce $\sqrt{729a^7b^5c^6d}$ to its simplest form.

Ans. $27a^3b^2c^3\sqrt{abd}$.

7. Reduce $\sqrt{675a^7b^5c^2d}$ to its simplest form.

Ans. $15a^3b^2c\sqrt{3abd}$.

8. Reduce $\sqrt{1445a^3c^8d^4}$ to its simplest form.

Ans. $17ac^4d^2\sqrt{5a}$.

9. Reduce $\sqrt{1008a^9d^7m^8}$ to its simplest form.

Ans. $12a^4d^3m^4\sqrt{7ad}$.

10. Reduce $\sqrt{2156a^{10}b^8c^6}$ to its simplest form.

Ans. $14a^5b^4c^3\sqrt{11}$.

11. Reduce $\sqrt{405a^7b^6d^8}$ to its simplest form.

Ans. $9a^3b^3d^4\sqrt{5a}$

106. Since like signs in both the factors give a plus sign in the product, the square of $-a$, as well as that of $+a$, will be a^2; hence the root of a^2 is either $+a$ or $-a$ Also, the square root of $25a^2b^4$ is either $+5ab^2$ or $-5ab^2$. Whence we may conclude, that if a monomial is positive its square root may be affected either with the sign $+$ or $-$; thus, $\sqrt{9a^4}=\pm 3a^2$; for, $+3a^2$ or $-3a^2$, squared, gives $9a^4$. The double sign $\pm$ with which the root is affected is read *plus or minus.*

If the proposed monomial were *negative*, it would be impossible to extract its root, since it has just been shown that the square of every quantity, whether positive or negative, is essentially positive. Therefore,

$$\sqrt{-9}, \quad \sqrt{-4a^2}, \quad \sqrt{-8a^2b},$$

are algebraic symbols which indicate operations that cannot be performed. They are called *imaginary quantities*, or rather *imaginary expressions*, and are frequently met with in the resolution of equations of the second degree. These symbols can, however, by extending the rules, be simplified in the same manner as those irrational expressions which indicate operations that cannot be performed. Thus, $\sqrt{-9}$ may be reduced by (Art. **104**). Thus,

$$\sqrt{-9}=\sqrt{9}\times\sqrt{-1}=3\sqrt{-1},$$

and $$\sqrt{-4a^2}=\sqrt{4a^2}\times\sqrt{-1}=2a\sqrt{-1}:$$ also,

$$\sqrt{-8a^2b}=\sqrt{4a^2\times -2b}=2a\sqrt{-2b}=2a\sqrt{2b}\times\sqrt{-1}.$$

QUEST.—**106**. What sign is placed before the square root of a monomial? Why may you place the sign plus or minus? What is an imaginary quantity? Why is it called imaginary?

Of the Calculus of Radicals of the Second Degree

107. A *radical quantity* is the indicated root of an imperfect power.

The extraction of the square root gives rise to such expressions as $\sqrt{a}$, $3\sqrt{b}$, $7\sqrt{2}$, which are called *irrational quantities*, or *radicals of the second degree.* We will now establish rules for performing the four fundamental operations on these expressions.

108. Two radicals of the second degree are *similar*, when the quantities under the radical sign are the same in both. Thus, $3\sqrt{b}$ and $5c\sqrt{b}$ are similar radicals; and so also are $9\sqrt{2}$ and $7\sqrt{2}$.

Addition.

109. Radicals of the second degree may be added together by the following

RULE.

I. *If the radicals are similar add their coefficients, and to the sum annex the common radical.*

II. *If the radicals are not similar, connect them together with their proper signs.*

Thus, $3a\sqrt{b}+5c\sqrt{b}=(3a+5c)\sqrt{b}$.

QUEST.—**107**. What is a radical quantity? What are such quantities called?—**108**. When are radicals of the second degree similar?—**109**. How do you add similar radicals of the second degree? How do you add radicals which are not similar?

In like manner.

$$7\sqrt{2a}+3\sqrt{2a}=(7+3)\sqrt{2a}=10\sqrt{2a}.$$

Two radicals, which do not appear to be similar at first sight, may become so by simplification (Art. **104**).

For example,

$$\sqrt{48ab^2}+b\sqrt{75a}=4b\sqrt{3a}+5b\sqrt{3a}=9b\sqrt{3a};$$

and $$2\sqrt{45}+3\sqrt{5}=6\sqrt{5}+3\sqrt{5}=9\sqrt{5}.$$

When the radicals are not similar, the addition or subtraction can only be indicated. Thus, in order to add $3\sqrt{b}$ to $5\sqrt{a}$, we write

$$5\sqrt{a}+3\sqrt{b}$$

EXAMPLES.

1. What is the sum of $\sqrt{27a^2}$ and $\sqrt{48a^2}$?

Ans. $7a\sqrt{3}$.

2. What is the sum of $\sqrt{50a^4b^2}$ and $\sqrt{72a^4b^2}$?

Ans. $11a^2b\sqrt{2}$.

3. What is the sum of $\sqrt{\frac{3a^2}{5}}$ and $\sqrt{\frac{a^2}{15}}$?

Ans. $4a\sqrt{\frac{1}{15}}$.

4. What is the sum of $\sqrt{125}$ and $\sqrt{500a^2}$?

Ans. $(5+10a)\sqrt{5}$.

5. What is the sum of $\sqrt{\frac{50}{147}}$ and $\sqrt{\frac{100}{294}}$?

Ans. $\frac{10}{21}\sqrt{6}$.

6. What is the sum of $\sqrt{98a^2x}$ and $\sqrt{36x^2-36a^2}$?

Ans. $7a\sqrt{2x}+6\sqrt{x^2-a^2}$.

7. What is the sum of $\sqrt{98a^2x}$ and $\sqrt{288a^4x^5}$?

Ans. $(7a+12a^2x^2)\sqrt{2x}$.

8. Required the sum of $\sqrt{72}$ and $\sqrt{128}$.

Ans. $14\sqrt{2}$.

9. Required the sum of $\sqrt{27}$ and $\sqrt{147}$.

Ans. $10\sqrt{3}$.

10. Required the sum of $\sqrt{\frac{2}{3}}$ and $\sqrt{\frac{27}{50}}$.

Ans. $\frac{19}{30}\sqrt{6}$.

11. Required the sum of $2\sqrt{a^2b}$ and $3\sqrt{64bx^4}$.

Ans. $(2a+24x^2)\sqrt{b}$.

12. Required the sum of $\sqrt{243}$ and $10\sqrt{363}$.

Ans. $119\sqrt{3}$.

13. What is the sum of $\sqrt{320a^2b^2}$ and $\sqrt{245a^8b^6}$?

Ans. $(8ab+7a^4b^3)\sqrt{5}$.

14 What is the sum of $\sqrt{75a^6b^7}$ and $\sqrt{300a^6b^5}$?

Ans. $(5a^3b^3+10a^3b^2)\sqrt{3b}$.

Subtraction.

110. To subtract one radical expression from another, we have the following

RULE.

I. *If the radicals are similar, subtract their coefficients, and to the difference annex the common radical.*

II. *If the radicals are not similar, their difference can only be indicated by the minus sign.*

EXAMPLES.

1. What is the difference between $3a\sqrt{b}$ and $a\sqrt{b}$?

Here $3a\sqrt{b}-a\sqrt{b}=2a\sqrt{b}$. *Ans.*

2. From $9a\sqrt{27b^2}$ subtract $6a\sqrt{27b^2}$.

First, $9a\sqrt{27b^2}=27ab\sqrt{3}$, and $6a\sqrt{27b^2}=18ab\sqrt{3}$;

and $27ab\sqrt{3}-18ab\sqrt{3}=9ab\sqrt{3}$ *Ans.*

3 What is the difference of $\sqrt{75}$ and $\sqrt{48}$?

Ans. $\sqrt{3}$.

4. What is the difference of $\sqrt{24a^2b^2}$ and $\sqrt{54b^4}$?

Ans. $(2ab-3b^2)\sqrt{6}$

QUEST.—**110.** How do you subtract similar radicals? How do you subtract radicals which are not similar?

5. Required the difference of $\sqrt{\frac{3}{5}}$ and $\sqrt{\frac{5}{27}}$.

Ans. $\frac{4}{45}\sqrt{15}$.

6. What is the difference of $\sqrt{128a^3b^2}$ and $\sqrt{32a^9}$?

Ans. $(8ab-4a^4)\sqrt{2a}$.

7. What is the difference of $\sqrt{48a^3b^3}$ and $\sqrt{9ab}$?

Ans. $4ab\sqrt{3ab}-3\sqrt{ab}$.

8. What is the difference of $\sqrt{242a^5b^5}$ and $\sqrt{2a^3b^3}$?

Ans. $(11a^2b^2-ab)\sqrt{2ab}$.

9. What is the difference of $\sqrt{\frac{3}{4}}$ and $\sqrt{\frac{3}{9}}$?

Ans. $\frac{1}{6}\sqrt{3}$.

10. What is the difference of $\sqrt{320a^2}$ and $\sqrt{80a^2}$?

Ans. $4a\sqrt{5}$.

11 What is the difference between

$\sqrt{720a^3b^3}$ and $\sqrt{245abc^2d^2}$?

Ans. $(12ab-7cd)\sqrt{5ab}$.

12. What is the difference between

$\sqrt{968a^2b^2}$ and $\sqrt{200a^2b^2}$?

Ans. $12ab\sqrt{2}$.

13. What is the difference between

$\sqrt{112a^8b^6}$ and $\sqrt{28a^8b^6}$?

Ans. $2a^4b^3\sqrt{7}$.

Multiplication.

111. For the multiplication of radicals, we have the following

RULE.

I. *Multiply the quantities under the radical signs together, and place the common radical over the product.*

II. *If the radicals have coefficients, we multiply them together, and place the product before the common radical.*

Thus, $\sqrt{a} \times \sqrt{b} = \sqrt{ab}$;

This is the principle of Art. **104**, taken in the inverse order.

EXAMPLES.

1. What is the product of $3\sqrt{5ab}$ and $4\sqrt{20a}$?

Ans. $120a\sqrt{b}$.

2. What is the product of $2a\sqrt{bc}$ and $3a\sqrt{bc}$?

Ans. $6a^2bc$

3. What is the product of $2a\sqrt{a^2+b^2}$ and $-3a\sqrt{a^2+b^2}$?

Ans. $-6a^2(a^2+b^2)$.

Quest.—**111.** How do you multiply quantities which are under radical signs? When the radicals have coefficients, now do you multiply them?

4. What is the product of $3\sqrt{2}$ and $2\sqrt{8}$?

Ans. 24.

5. What is the product of $\frac{5}{3}\sqrt{\frac{3}{8}a^2b}$ and $\frac{2}{10}\sqrt{\frac{2}{5}c^2b}$?

Ans. $\frac{1}{30}abc\sqrt{15}$.

6. What is the product of $2x+\sqrt{b}$ and $2x-\sqrt{b}$?

Ans. $4x^2-b$.

7. What is the product of

$$\sqrt{a+2\sqrt{b}} \quad \text{and} \quad \sqrt{a-2\sqrt{b}}\,?$$

Ans. $\sqrt{a^2-4b}$.

8. What is the product of $3a\sqrt{27a^3}$ by $\sqrt{2a}$?

Ans. $9a^3\sqrt{6}$.

Division.

112. To divide one radical by another, we have the following

RULE.

I. *Divide one of the quantities under the radical sign by the other, and place the common radical over the quotient.*

II. *If the radicals have coefficients, divide the coefficient of the dividend by the coefficient of the divisor, and place the quotient before the common radical.*

QUEST.—**112.** How do you divide quantities which are under the radical sign? When the radicals have coefficients, how do you divide them?

Thus, $\frac{\sqrt{a}}{\sqrt{b}}=\sqrt{\frac{a}{b}}$; for the squares of these two expressions are equal to the same quantity $\frac{a}{b}$; hence the expressions themselves must be equal.

EXAMPLES.

1. Divide $5a\sqrt{b}$ by $2b\sqrt{c}$. *Ans.* $\frac{5a}{2b}\sqrt{\frac{b}{c}}$.

2. Divide $12ac\sqrt{6bc}$ by $4c\sqrt{2b}$. *Ans.* $3a\sqrt{3c}$.

3. Divide $6a\sqrt{96b^4}$ by $3\sqrt{8b^2}$. *Ans.* $4ab\sqrt{3}$.

4. Divide $4a^2\sqrt{50b^5}$ by $2a^2\sqrt{5b}$. *Ans.* $2b^2\sqrt{10}$.

5. Divide $26a^3b\sqrt{81a^2b^2}$ by $13a\sqrt{9ab}$. *Ans.* $6a^2b\sqrt{ab}$.

6. Divide $84a^3b^4\sqrt{27ac}$ by $42ab\sqrt{3a}$. *Ans.* $6a^2b^3\sqrt{c}$.

7. Divide $\sqrt{\frac{1}{8}a^2}$ by $\sqrt{2}$. *Ans.* $\frac{1}{4}a$.

8. Divide $6a^2b^2\sqrt{20a^3}$ by $12\sqrt{5a}$. *Ans.* a^3b^2.

9. Divide $6a\sqrt{10b^2}$ by $3\sqrt{5}$. *Ans.* $2ab\sqrt{2}$.

10. Divide $48b^4\sqrt{15}$ by $2b^2\sqrt{\frac{1}{15}}$. *Ans.* $360b^7$.

11. Divide $8a^2b^4c^3\sqrt{7d^3}$ by $2a\sqrt{28d}$. *Ans.* $2ab^4c^3d$.

12. Divide $96a^4c^3\sqrt{98b^5}$ by $48abc\sqrt{2b}$. *Ans.* $14a^3bc^2$.

13. Divide $27a^5b^6\sqrt{21a^3}$ by $\sqrt{7a}$.

Ans. $27a^6b^6\sqrt{3}$.

14. Divide $18a^8b^6\sqrt{8a^4}$ by $6ab\sqrt{a^2}$.

Ans. $6a^8b^5\sqrt{2}$

To Extract the Square Root of a Polynomial.

113. Before explaining the rule for the extraction of the square root of a polynomial, let us first examine the squares of several polynomials: we have

$$(a+b)^2=a^2+2ab+b^2,$$

$$(a+b+c)^2=a^2+2ab+b^2+2(a+b)c+c^2,$$

$$(a+b+c+d)^2=a^2+2ab+b^2+2(a+b)c+c^2$$
$$+2(a+b+c)d+d^2.$$

The *law* by which these squares are formed can be enunciated thus:

The square of any polynomial contains the square of the first term, plus twice the product of the first term by the second, plus the square of the second; plus twice the first two terms multiplied by the third, plus the square of the third; plus twice the first three terms multiplied by the fourth, plus the square of the fourth; and so on.

QUEST.—**113.** What is the square of a binomial equal to? What is the square of a trinomial equal to? What is the square of any polynomial equal to?

114. Hence, to extract the square root of a polynomial we have the following

RULE.

I. *Arrange the polynomial with reference to one of its letters and extract the square root of the first term: this will give the first term of the root.*

II. *Divide the second term of the polynomial by double the first term of the root, and the quotient will be the second term of the root.*

III. *Then form the square of the two terms of the root found, and subtract it from the first polynomial, and then divide the first term of the remainder by double the first term of the root, and the quotient will be the third term.*

IV. *Form the double products of the first and second terms, by the third, plus the square of the third; then subtract all these products from the last remainder, and divide the first term of the result by double the first term of the root, and the quotient will be the fourth term. Then proceed in the same manner to find the other terms.*

EXAMPLES.

1 Extract the square root of the polynomial

$$49a^2b^2-24ab^3+25a^4-30a^3b+16b^4.$$

First arrange it with reference to the letter a.

$25a^4-30a^3b+49a^2b^2-24ab^3+16b^4$	$5a^2-3ab+4b^2$
$25a^4-30a^3b+\ 9a^2b^2$	$10a^2$
$40a^2b^2-24ab^3+16b^4$	1st Rem.
$40a^2b^2-24ab^3+16b^4$	
0 . . .	2d Rem.

After having arranged the polynomial with reference to a, extract the square root of $25a^4$, this gives $5a^2$, which is placed at the right of the polynomial; then divide the second term, $-30a^3b$, by the double of $5a^2$, or $10a^2$; the quotient is $-3ab$, and is placed at the right of $5a^2$. Hence, the first two terms of the root are $5a^2-3ab$. Squaring this binomial, it becomes $25a^4-30a^3b+9a^2b^2$, which, subtracted from the proposed polynomial, gives a remainder, of which the first term is $40a^2b^2$. Dividing this first term by $10a^2$, (the double of $5a^2$), the quotient is $+4b^2$; this is the third term of the root, and is written on the right of the first two terms. By forming the double product of $5a^2-3ab$ by $4b^2$, and at the same time squaring $4b^2$, we find the polynomial $40a^2b^2-24ab^3+16b^4$, which, subtracted from the first remainder, gives 0. Therefore $5a^2-3ab+4b^2$ is the required root.

2. Find the square root of $a^4+4a^3x+6a^2x^2+4ax^3+x^4$.

Ans. $a^2+2ax+x^2$.

3. Find the square root of $a^4-4a^3x+6a^2x^2-4ax^3+x^4$.

Ans. $a^2-2ax+x^2$.

4. Find the square root of

$$4x^6+12x^5+5x^4-2x^3+7x^2-2x+1.$$

Ans. $2x^3+3x^2-x+1$.

5. Find the square root of

$$9a^4-12a^3b+28a^2b^2-16ab^3+16b^4.$$

Ans. $3a^2-2ab+4b^2$.

QUEST.—**114.** Give the rule for extracting the square root of a polynomial? What is the first step? What the second? What the third? What the fourth?

6. What is the square root of

$$x^4-4ax^3+4a^2x^2-4x^2+8ax+4.$$

Ans. $x^2-2ax-2.$

7. What is the square root of

$$9x^2-12x+6xy+y^2-4y+4.$$

Ans. $3x+y-2.$

8. What is the square root of $y^4-2y^2x^2+2x^2-2y^2+1+x^4$.

Ans. $y^2-x^2-1.$

9. What is the square root of $9a^4b^4-30a^3b^3+25a^2b^2$?

Ans. $3a^2b^2-5ab$

10. Find the square root of

$$25a^4b^2-40a^3b^2c+76a^2b^2c^2-48ab^2c^3+36b^2c^4-30a^4bc$$
$$+24a^3bc^2-36a^2bc^3+9a^4c^2.$$

Ans. $5a^2b-3a^2c-4abc+6bc^2.$

115. We will conclude this subject with the following remarks.

1st. A binomial can never be a perfect square, since we know that the square of the most simple polynomial, viz: a binomial, contains three distinct parts, which cannot experience any reduction amongst themselves. Thus, the expression a^2+b^2 is not a perfect square; it wants the term $\pm 2ab$ in order that it should be the square of $a\pm b$.

2nd. In order that a trinomial, when arranged, may be a perfect square, its two extreme terms must be squares, and the middle term must be the double product of the square roots of the two others. Therefore, to obtain the square root of a trinomial when it is a perfect square; *Extract the roots of the two extreme terms, and give these roots the same or contrary signs, according as the middle term is positive or* negative. *To verify it, see if the double product of the two roots gives the middle term of the trinomial.* Thus,

$$9a^6-48a^4b^2+64a^2b^4 \quad \text{is a perfect square,}$$

since $\sqrt{9a^6}=3a^3$, and $\sqrt{64a^2b^4}=-8ab^2$,

and also $2\times3a^3\times-8ab^2=-48a^4b^2=$ the middle term.

But $4a^2+14ab+9b^2$ is not a perfect square: for although $4a^2$ and $+9b^2$ are the squares of $2a$ and $3b$, yet $2\times2a\times3b$ is not equal to $14ab$.

3rd. In the series of operations required in a general example, when the first term of one of the remainders is not exactly divisible by twice the first term of the root, we may conclude that the proposed polynomial is not a perfect square. This is an evident consequence of the course of reasoning, by which we have arrived at the general rule for extracting the square root.

4th. When the polynomial is *not a perfect square*, it may be simplified (See Art. **104**.)

Take, for example, the expression $\sqrt{a^3b+4a^2b^2+4ab^3}$.

The quantity under the radical is not a perfect square; but it can be put under the form $ab(a^2+4ab+4b^2)$. Now, the factor between the parenthesis is evidently the square of $a+2b$, whence we may conclude that,

$$\sqrt{a^3b+4a^2b^2+4ab^3}=(a+2b)\sqrt{ab}.$$

2. Reduce $\sqrt{2a^2b-4ab^2+2b^3}$ to its simple form.

Ans. $(a-b)\sqrt{2b}$.

QUEST.—**115**. Can a binomial ever be a perfect power? Why not? When is a trinomial a perfect square? When, in extracting the square root we find that the first term of the remainder is not divisible by twice the root, is the polynomial a perfect power or not?

CHAPTER VI.

Equations of the Second Degree.

116. An Equation of the second degree is one in which the greatest exponent of the unknown quantity is equal to 2.

If the equation contains two unknown quantities, it is of the second degree when the greatest sum of the exponents with which the unknown quantity is affected, in any term, is equal to 2. Thus,

$$x^2=a, \quad ax^2+bx=c, \quad \text{and} \quad xy+x=d^2,$$

are equations of the second degree.

117. Equations of the second degree are divided into two classes:

1st. Equations which involve only the square of the unknown quantity and known terms. These are called, *Incomplete Equations.*

2d. Equations which involve the first and second powers of the unknown quantity, and known terms. These are called, *Complete Equations.*

QUEST.—**116.** What is an equation of the second degree?—**117.** Into how many classes are equations of the second degree divided? What is an incomplete equation? What is a complete equation?

Thus, $x^2+2x^2-5=7$

and $5x^2-3x^2-4=a$

are incomplete equations: and

$$3x^2-5x-3x^2+a=b$$
$$2x^2-8x^2-x-c=d$$

are complete equations.

Of Incomplete Equations.

118. If we take an incomplete equation of the form

$$14x^2-8x^2=40-2x^2$$

we have, by collecting the coefficients of x^2,

$$8x^2=40, \text{ or } x^2=5.$$

Again,—if we have the equation

$$ax^2+bx^2+d=f,$$

we shall have,

$$(a+b)x^2=f-d, \text{ and } x^2=\frac{f-d}{a+b}=m,$$

by substituting m for the known terms which compose the second member. Hence,

Every incomplete equation can be reduced to an equation involving two terms, of the form

$$x^2=m,$$

and from this circumstance the incomplete equations are often called *equations involving two terms.*

From which we have, by extracting the square root of both members, $x=\sqrt{m}.$

QUEST.—118. To what form may every incomplete equation be reduced? What are incomplete equations often called?

1. What number is that which being multiplied by itself the product will be 144.

Let $x=$ the number: then

$$x \times x = x^2 = 144.$$

It is plain that the value of x will be found by extracting the square root of both members of the equation: that is

$$\sqrt{x^2} = \sqrt{144}: \text{ that is, } x = 12.$$

2. A person being asked how much money he had, said if the number of dollars be squared and 6 be added, the sum will be 42: How much had he?

Let $x=$ the number of dollars.

Then by the conditions

$$x^2 + 6 = 42:$$

hence, $$x^2 = 42 - 6 = 36$$

and $$x = 6.$$

Ans. \$6.

3. A grocer being asked how much sugar he had sold to a person, answered, if the square of the number of pounds be multiplied by 7, the product will be 1575. How many pounds had he sold?

Denote the number of pounds by x.

Then by the conditions of the question

$$7x^2 = 1575:$$

hence, $$x^2 = 225$$

and $$x = 15.$$

Ans. 15

4. A person being asked his age said, if from the square of my age you take 192, the remainder will be the square of half my age: what was his age?

Denote his age by x.

Then, by the conditions of the question

$$x^2-192=\left(\frac{1}{2}x\right)^2=\frac{x^2}{4},$$

and by clearing the fractions

$$4x^2-768=x^2;$$

hence, $$4x^2-x^2=768,$$

and $$3x^2=768$$

$$x^2=256$$

$$x=16.$$

Ans. 16.

5. What number is that whose eighth part multiplied by its fifth part and the product divided by 4, shall give a quotient equal to 40?

Let $x=$ the number,

By the conditions of the question

$$\left(\frac{1}{8}x\times\frac{1}{5}x\right)\div 4=40,$$

hence, $$\frac{x^2}{160}=40$$

by clearing fractions,

$$x^2=6400$$
$$x=80.$$

Ans. 80.

119. Hence, to find the value of x we have the following

RULE.

1. *Find the value of* $\mathbf{x}^2$; *and then extract the square root of both members of the equation.*

4. What is the value of x in the equation

$$3x^2+8=5x^2-10.$$

By transposition $3x^2-5x^2=-10-8$,

by reducing $-2x^2=-18$,

by dividing by 2 and changing the signs

$$x^2=9,$$

by extracting the square root $x=3$.

We should, however, remark that the square root of 9, is either $+3$, or -3. For,

$$+3\times+3=9 \quad \text{and} \quad -3\times-3=9.$$

Hence, when we have the equation

$$x^2=9,$$

we have $x=+3$ and $x=-3$.

120. A *root* of an equation is any expression which being substituted for the unknown quantity, will satisfy the equation, that is, render the two members equal to each other. Thus, in the equation

$$x^2=9$$

there are two roots, $+3$ and -3; for either of these numbers being substituted for x will satisfy the equation.

7. Again, if we take the equation

$$x^2=m,$$

we shall have

$$x=+\sqrt{m}, \text{ and } \quad x=-\sqrt{m}$$

For, $\quad (+\sqrt{m})^2=m;$

and $\quad (-\sqrt{m})^2=m;$

Hence we may conclude,

1st. *That every incomplete equation of the second degree has two roots.*

2d. *That these roots are numerically equal, but have contrary signs.*

8. What are the roots of the equation

$$3x^2+6=4x^2-10.$$

Ans. $x=+4$ and $x=-4.$

9. What are the roots of the equation

$$\frac{1}{3}x^2-8=\frac{x^2}{9}+10.$$

Ans. $x=+9$ and $x=-9.$

10. What are the roots of the equation

$$4x^2+13-2x^2=45.$$

Ans. $x=+4$ and $x=-4$

QUEST.—119. How do you resolve an incomplete equation? 120. What is the root of an equation? What are the roots of the equation $x^2=9$? Of the equation $x^2=m$? How many roots has every incomplete equation? How do those roots compare with each other?

8. What are the roots of the equation

$$6x^2-7=3x^2+5.$$

Ans. $x=+2,\quad x=-2$

9. What are the roots of the equation.

$$8+5x^2=\frac{x^2}{5}+4x^2+28.$$

Ans. $x=+5,\quad x=-5$

10. Find a number such that one-third of it multiplied by one-fourth shall be equal to 108 ?

Ans. 36.

11. What number is that whose sixth part multiplied by its fifth part and product divided by ten, shall give a quotient equal to 3 ?

Ans. 30.

12. What number is that whose square, plus 18, shall be equal to half its square plus $30\frac{1}{2}$.

Ans. 5.

13. What numbers are those which are to each other as 1 to 2 and the difference of whose squares is equal to 75.

Let $x=$ the less number.

Then $2x=$ the greater.

Then by the conditions of the question

$$4x^2-x^2=75,$$

hence, $3x^2=75$;

and by dividing by 3, $x^2=25$ and $x=5$,

and $2x=10$.

Ans. 5 and 10.

14. What two numbers are those which are to each other as 5 to 6, and the difference of whose squares is 44.

Let $x=$ the greatest number.

Then $\frac{5}{6}x=$ the least.

By the conditions of the question

$$x^2-\frac{25}{36}x^2=44.$$

by clearing fractions,

$$36x^2-25x^2=1584;$$

hence, $$11x^2=1584,$$

and $$x^2=144,$$

hence, $$x=12,$$

and $$\frac{5}{6}x=10.$$

Ans. 10 and 12.

15. What two numbers are those which are to each other as 3 to 4, and the difference of whose squares is 28?

Ans. 6 and 8.

16. What two numbers are those which are to each other as 5 to 11, and the sum of whose square is 584?

Ans. 10 and 22.

17. A says to B, my son's age is one quarter of yours, and the difference between the squares of the numbers representing their ages is 240: what were their ages?

Ans. { Eldest 16. Younger 4.

When there are two unknown quantities

121. When there are two or more unknown quantities, *eliminate one of them by the rule of Article* **77** : *there will thus arise a new equation with but a single unknown quantity, the value of which may be found by the rule already given.*

1. There is a room of such dimensions, that the difference of the sides multiplied by the less is equal to 36, and the product of the sides is equal to 360 : what are the sides ?

Let $x=$ the less side ;
$y=$ the greater.

Then, by the 1st condition,

$$(y-x)x=36\ ;$$

and by the 2nd, $$xy=360.$$

From the first equation, we have

$$xy-x^2=36\ ;$$

and by subtraction, $$x^2=324.$$

Hence, $$x=\sqrt{324}=18\ ;$$

$$y=\frac{360}{18}=20.$$

Ans. $x=18, y=20$.

QUEST.—121. How do you resolve the equation when there are two or more unknown quantities ?

2. A merchant sells two pieces of muslin, which together measure 12 yards. He received for each piece just so many dollars per yard as the piece contained yards. Now, he gets four times as much for one piece as for the other: how many yards in each piece?

Let $x=$ the number in the larger piece;

$y=$ the number in the shorter piece.

Then, by the conditions of the question,

$$x+y=12.$$

$x\times x=x^2=$ what he got for the larger piece;

$y\times y=y^2=$ what he got for the shorter.

And $x^2=4y^2$, by the 2nd condition.

$x=2y$, by extracting the square root

Substituting this value of x in the first equation, we have

$$y+2y=12;$$

and consequently, $y=4$,

and $x=8$.

Ans. 8 and 4.

3. What two numbers are those whose product is 30, and quotient $3\frac{1}{3}$? *Ans.* 10 and 3.

4. The product of two numbers is a, and their quotient b: what are the numbers?

Ans. $\sqrt{ab}$ and $\sqrt{\frac{a}{b}}$

5. The sum of the squares of two numbers is 117, an the difference of their squares 45: what are the numbers? *Ans.* 9 and 6

6. The sum of the squares of two numbers is a, and the difference of their squares is b: what are the numbers?

Ans. $x=\sqrt{\frac{a+b}{2}}$, $y=\sqrt{\frac{a-b}{2}}$.

7. What two numbers are those which are to each other as 3 to 4, and the sum of whose squares is 225?

Ans. 9 and 12.

8. What two numbers are those which are to each other as m to n, and the sum of whose squares is equal to a^2?

Ans. $\frac{ma}{\sqrt{m^2+n^2}}$, $\frac{na}{\sqrt{m^2+n^2}}$.

9. What two numbers are those which are to each other as 1 to 2, and the difference of whose squares is 75?

Ans. 5 and 10.

10. What two numbers are those which are to each other as m to n, and the difference of whose squares is equal to b^2?

Ans. $\frac{mb}{\sqrt{m^2-n^2}}$, $\frac{nb}{\sqrt{m^2-n^2}}$.

11. A certain sum of money is placed at interest for six months, at 8 per cent. per annum. Now, if the amount be multiplied by the number expressing the interest, the product will be 562500: what is the amount at interest?

Ans. $3750.

12. A person distributes a sum of money between a number of women and boys. The number of women is to the number of boys as 3 to 4. Now, the boys receive one-half as many dollars as there are persons, and the women twice as many dollars as there are boys, and together they receive 138 dollars: how many women were there, and how many boys?

36 women
48 boys.

Of Complete Equations.

122. We have already seen (Art. **117**), that a complete equation of the second degree, contains the square of the unknown quantity, the first power of the unknown quantity, and known terms.

1. If we have the complete equation

$$5x^2-2x^2+8=9x+32,$$

we have, by transposing and reducing,

$$3x^2-9x=24,$$

and by dividing by 3,

$$x^2-3x=8,$$

an equation containing but three terms.

2. If we have the equation

$$a^2x^2+3abx+x^2=cx+d,$$

by collecting the coefficients of x^2 and x, we have

$$(a^2+1)x^2+(3ab-c)x=d;$$

and dividing by the coefficient of x^2, we have

$$x^2+\frac{3ab-c}{a^2+1}x=\frac{d}{a^2+1}.$$

QUEST.—122. How many terms does a complete equation of the second degree contain? Of what is the first term composed? The second? The third?

If we represent the coefficient of x by $2p$, and the known term by q, we have

$$x^2+2px=q,$$

an equation containing but three terms.

Hence, *we see that every complete equation of the second degree can be reduced to an equation containing but three terms*

123. We wish now to show that there are four forms under which this equation will be expressed, each depending on the signs of $2p$ and q.

1st. Let us for the sake of illustration, make

$$2p=+4, \quad \text{and} \quad q=+5:$$

we shall then have $\quad x^2+4x=5.$

2nd. Let us now suppose

$$2p=-4, \quad \text{and} \quad q=+5:$$

we shall then have $\quad x^2-4x=5.$

3rd. If we make

$$2p=+4, \quad \text{and} \quad q=-5,$$

we have $\quad x^2+4x=-5.$

4th. If we make

$$2p=-4, \quad \text{and} \quad q=-5,$$

we have $\quad x^2-4x=-5.$

QUEST.—**123**. Under how many forms may every equation of the second degree be expressed? On what will these forms depend? What are the signs of the coefficient of x and the known term, in the first form? What in the second? What in the third? What in the fourth? Repeat the four forms.

We therefore conclude that every complete equation of the second degree may be reduced to one of these forms :

$$x^2+2px=+q,\quad \text{1st form.}$$
$$x^2-2px=+q,\quad \text{2nd form.}$$
$$x^2+2px=-q,\quad \text{3rd form.}$$
$$x^2-2px=-q,\quad \text{4th form.}$$

124. REMARK.—If, in reducing an equation to either of these forms, the second power of the unknown quantity should have a negative sign, it must be rendered positive by changing the sign of every term of the equation.

125. We are next to show the manner in which the value of the unknown quantity may be found. We have seen (Art. **38**), that

$$(x+p)^2=x^2+2px+p^2\ ;$$

and comparing this square with the first and third forms, we see that the first member in each contains two terms of the square of a binomial, viz : the square of the first term plus twice the product of the 2nd term by the first. If, then, we take half the coefficient of x, viz : p, and square it, and add to both members, the equations take the form

$$x^2+2px+p^2=q+p^2,$$
$$x^2+2px+p^2=-q+p^2,$$

in which the first members are perfect squares. This is

QUEST.—**124**. If in reducing an equation to either of these forms the coefficient of x^2 is negative, what do you do ?—**125**. What is the square of a binomial equal to ? What does the first member in each form contain ? How do you render the first member a perfect square ? What is this called ?

called completing the square. Then, by extracting the square root of both members of the equation, we have

$$x+p=\pm\sqrt{q+p^2},$$

and $$x+p=\pm\sqrt{-q+p^2},$$

which gives, by transposing p,

$$x=-p\pm\sqrt{q+p^2},$$

$$x=-p\pm\sqrt{-q+p^2}.$$

126. If we compare the second and fourth forms with the square

$$(x-p)^2=x^2-2px+p^2,$$

we also see that half the coefficient of x being squared and added to both members, will make the first members perfect squares. Having made the additions, we have

$$x^2-2px+p^2=q+p^2,$$

$$x^2-2px+p^2=-q+p^2.$$

Then, by extracting the square root of both members we have

$$x-p=\pm\sqrt{q+p^2},$$

and $$x-p=\pm\sqrt{-q+p^2};$$

and by transposing $-p$, we find

$$x=p\pm\sqrt{q+p^2},$$

and $$x=p\pm\sqrt{-q+p^2}.$$

QUEST.—**126.** In the second form, how do you make the first member a perfect square?

127. Hence, for the resolution of every equation of the econd degree, we have the following

RULE.

I. *Reduce the equation to one of the known forms.*

II. *Take half the coefficient of the second term, square it, and add the result to both members of the equation.*

III. *Then extract the square root of both members of the equation; after which, transpose the known term to the second member.*

REMARK.—The square root of the first member is always equal to the square root of the first term, plus or minus half the coefficient of x.

EXAMPLES IN THE FIRST FORM.

1. What are the values of x in the equation

$$2x^2+8x=64\ ?$$

If we first divide by the coefficient 2, we obtain

$$x^2+4x=32.$$

Then, completing the square,

$$x^2+4x+4=32+4=36.$$

Extracting the root,

$$x+2=\pm\sqrt{36}=+6 \text{ or } -6.$$

Hence, $\qquad x=-2+6=+4;$

or, $\qquad x=-2-6=-8.$

QUEST.—**127.** Give the general rule for resolving an equation of the second degree. What is the first step? What the second? What the third? What is the square root of the first member always equal to?

That is, in this form the smaller root is positive, and the larger negative.

Verification.

If we take the positive value, viz: $x=+4$,

he equation $x^2+4x=32$

gives $4^2+4\times4=32$:

and if we take the negative value of x, viz: $x=-8$,

the equation $x^2+4x=32$

gives $(-8)^2+4(-8)=64-32=32$.

From which we see that either of the values of x, viz: $x=+4$ or $x=-8$, will satisfy the equation.

2. What are the values of x in the equation

$$3x^2+12x-19=-x^2-12x+89\ ?$$

By transposing the terms, we have

$$3x^2+x^2+12x+12x=89+19:$$

and by reducing,

$$4x^2+24x=108\ ;$$

and dividing by the coefficient of x^2,

$$x^2+6x=27$$

Now, by completing the square,

$$x^2+6x+9=36\ ;$$

extracting the square root,

$$x+3=\pm\sqrt{36}=+6 \text{ or } -6:$$

hence, $x=+6-3=+3$;

or, $x=-6-3=-9$.

Verification.

If we take the plus root, the equation

$$x^2+6x=27$$

gives $$(3)^2+6(3)=27;$$

and for the negative root,

$$x^2+6x=27$$

gives $$(-9)^2+6(-9)=81-54=27.$$

4. What are the values of x in the equation

$$x^2-10x+15=\frac{x^2}{5}-34x+155.$$

By clearing the fractions, we have

$$5x^2-50x+75=x^2-170x+775:$$

by transposing and reducing, we obtain

$$4x^2+120x=700;$$

then, dividing by the coefficient of x^2, we have

$$x^2+30x=175;$$

and by completing the square,

$$x^2+30x+225=400;$$

and by extracting the square root,

$$x+15=\pm\sqrt{400}=+20 \text{ or } -20$$

Hence, $$x=+5 \text{ or } -35.$$

Verification.

For the plus value of x, the equation

$$x^2+30x=175$$

gives $$(5)^2+30\times5=25+150=175$$

And for the negative value of x, we have

$$(-35)^2+30(-35)=1225-1050=175.$$

5. What are the values of x in the equation

$$\frac{5}{6}x^2-\frac{1}{2}x+\frac{3}{4}=8-\frac{2}{3}x-x^2+\frac{273}{12}\ ?$$

Clearing the fractions, we have

$$10x^2-6x+9=96-8x-12x^2+273\ ;$$

transposing and reducing,

$$22x^2+2x=360\ ;$$

dividing both members by 22,

$$x^2+\frac{2}{22}x=\frac{360}{22}.$$

Add $\left(\frac{1}{22}\right)^2$ to both members, and the equation becomes

$$x^2+\frac{2}{22}x+\left(\frac{1}{22}\right)^2=\frac{360}{22}+\left(\frac{1}{22}\right)^2\ ;$$

whence, by extracting the square root,

$$x+\frac{1}{22}=\pm\sqrt{\frac{360}{22}+\left(\frac{1}{22}\right)^2},$$

therefore,

$$x=-\frac{1}{22}+\sqrt{\frac{360}{22}+\left(\frac{1}{22}\right)^2},$$

and $$x=-\frac{1}{22}-\sqrt{\frac{360}{22}+\left(\frac{1}{22}\right)^2}.$$

It remains to perform the numerical operations. In the first place, $\frac{360}{22}+\left(\frac{1}{22}\right)^2$ must be reduced to a single number, having $(22)^2$ for its denominator.

Now,
$$\frac{360}{22}+\left(\frac{1}{22}\right)^2=\frac{360\times 22+1}{(22)^2}=\frac{7921}{(22)^2};$$

extracting the square root of 7921, we find it to be 89; therefore,

$$\pm\sqrt{\frac{360}{22}+\left(\frac{1}{22}\right)^2}=\pm\frac{89}{22}.$$

Consequently, the plus value of x is

$$x=-\frac{1}{22}+\frac{89}{22}=\frac{88}{22}=4,$$

and the negative value is

$$x=-\frac{1}{22}-\frac{89}{22}=-\frac{45}{11};$$

that is, one of the two values of x which will satisfy the proposed equation is a positive whole number, and the other a negative fraction.

6. What are the values of x in the equation

$$3x^2+2x-9=76.$$

Ans. $\begin{cases} x=5 \\ x=-5\frac{2}{3}. \end{cases}$

7. What are the values of x in the equation

$$2x^2+8x+7=\frac{5x}{4}-\frac{x^2}{8}+197.$$

Ans. $\begin{cases} x=8 \\ x=-11\frac{3}{17}. \end{cases}$

8. What are the values of x in the equation

$$\frac{x^2}{4}-\frac{x}{3}+15=\frac{x^2}{9}-8x+95\tfrac{1}{4}.$$

Ans. $\begin{cases} x=9 \\ x=-64\tfrac{1}{5}. \end{cases}$

9. What are the values of x in the equation

$$\frac{x^2}{1}-\frac{5x}{4}-8=\frac{x}{2}-7x+6\tfrac{1}{2}.$$

Ans. $\begin{cases} x=2 \\ x=-7\tfrac{1}{4}. \end{cases}$

10. What are the values of x in the equation

$$\frac{x^2}{2}+\frac{x}{4}=\frac{x^2}{5}-\frac{x}{10}+\frac{13}{20}.$$

Ans. $\begin{cases} x=1 \\ x=-2\tfrac{1}{6} \end{cases}$

EXAMPLES IN THE SECOND FORM.

1. What are the values of x in the equation

$$x^2-8x+10=19.$$

By transposing,

$$x^2-8x=19-10=9,$$

then by completing the square

$$x^2-8x+16=9+16=25,$$

and by extracting the root

$$x-4=\pm\sqrt{25}=+5 \quad \text{or} \quad -5.$$

Hence,

$$x=4+5=9 \quad \text{or} \quad x=4-5=-1.$$

That is, in this form, the largest root is positive and the smaller negative.

Verification.

If we take the positive value of x, the equation

$$x^2-8x=9 \quad \text{gives} \quad (9)^2-8\times 9=81-72=9;$$

and if we take the negative value, the equation

$$x^2-8x=9 \quad \text{gives} \quad (-1)^2-8(-1)=1+8=9;$$

from which we see that both values alike satisfy the equation.

2. What are the values of x in the equation

$$\frac{x^2}{2}+\frac{x}{3}-15=\frac{x^2}{4}+x-14\tfrac{3}{4}.$$

By clearing the fractions, we have

$$6x^2+4x-180=3x^2+12x-177$$

and by transposing and reducing

$$3x^2-8x=3,$$

and dividing by the co-efficient of x^2, we obtain

$$x^2-\frac{8}{3}x=1.$$

Then, by completing the square, we have

$$x^2-\frac{8}{3}x+\frac{16}{9}=1+\frac{16}{9}=\frac{25}{9};$$

and by extracting the square root,

$$x-\frac{4}{3}=\pm\sqrt{\frac{25}{9}}=+\frac{5}{3} \quad \text{or} \quad -\frac{5}{3}.$$

Hence,

$$x=\frac{4}{3}+\frac{5}{3}=+3, \quad \text{or} \quad x=\frac{4}{3}-\frac{5}{3}=-\frac{1}{3}.$$

Verification.

For the positive value of x, the equation

$$x^2-\frac{8}{3}x=1$$

gives $$3^2-\frac{8}{3}\times 3=9-8=1:$$

and for the negative value, the equation

$$x^2-\frac{8}{3}x=1$$

gives $$\left(-\frac{1}{3}\right)^2-\frac{8}{3}\times-\frac{1}{3}=\frac{1}{9}+\frac{8}{9}=1.$$

3. What are the values of x in the equation

$$\frac{x^2}{2}-\frac{x}{3}+7\tfrac{3}{8}=8\ ?$$

Clearing the fractions, and dividing by the coefficient of x^2, we have

$$x^2-\frac{2}{3}x=1\tfrac{1}{4}.$$

Completing the square, we have

$$x^2-\frac{2}{3}x+\frac{1}{9}=1\tfrac{1}{4}+\frac{1}{9}=\frac{49}{36};$$

then, by extracting the square root, we have

$$x-\frac{1}{3}=\pm\sqrt{\frac{49}{36}}=+\frac{7}{6}\ \text{ or }\ -\frac{7}{6};$$

hence,

$$x=\frac{1}{3}+\frac{7}{6}=\frac{9}{6}=1\tfrac{1}{2},\ \text{ or }\ x=\frac{1}{3}-\frac{7}{6}=-\frac{5}{6}.$$

Verification

If we take the positive value of x, the equation

$$x^2-\frac{2}{3}x=1\tfrac{1}{4}$$

gives $$(1\tfrac{1}{2})^2-\frac{2}{3}\times 1\tfrac{1}{2}=2\tfrac{1}{4}-1=1\tfrac{1}{4}:$$

and for the negative value, the equation

$$x^2-\frac{2}{3}x=1\tfrac{1}{4}$$

gives $$\left(-\frac{5}{6}\right)^2-\frac{2}{3}\times-\frac{5}{6}=\frac{25}{36}+\frac{10}{18}=\frac{45}{36}=1\tfrac{1}{4}.$$

4. What are the values of x in the equation

$$4a^2-2x^2+2ax=18ab-18b^2\ ?$$

By transposing, changing the signs, and dividing by 2, it becomes

$$x^2-ax=2a^2-9ab+9b^2\ ,$$

whence, completing the square,

$$x^2-ax+\frac{a^2}{4}=\frac{9a^2}{4}-9ab+9b^2\ ;$$

extracting the square root,

$$x=\frac{a}{2}\pm\sqrt{\frac{9a^2}{4}-9ab+9b^2}.$$

Now, the square root of $\frac{9a^2}{4}-9ab+9b^2$, is evidently $\frac{3a}{2}-3b$. Therefore,

$$x=\frac{a}{2}\pm\left(\frac{3a}{2}-3b\right),\ \text{or}\ \begin{cases} x=\ \ 2a-3b, \\ x=-\ a+3b. \end{cases}$$

What will be the numerical values of x, if we suppose $a=6$ and $b=1$?

5. What are the values of x in the equation

$$\frac{1}{3}x-4-x^2+2x-\frac{4}{5}x^2=45-3x^2+4x\ ?$$

Ans. $\left\{\begin{array}{l} x=\ \ 7,12 \\ x=-5,73 \end{array}\right\}$ to within 0,01.

6. What are the values of x in the equation

$$8x^2-14x+10=2x+34\ ?$$

Ans. $\left\{\begin{array}{l} x=\ \ 3. \\ x=-1 \end{array}\right.$

7. What are the values of x in the equation

$$\frac{x^2}{4}-30+x=2x-22\ ?$$

Ans. $\left\{\begin{array}{l} x=\ \ 8. \\ x=-4. \end{array}\right.$

8. What are the values of x in the equation

$$x^2-3x+\frac{x^2}{2}=9x+13\tfrac{1}{2}\ ?$$

Ans. $\left\{\begin{array}{l} x=\ \ 9 \\ x=-1 \end{array}\right.$

9. What are the values of x in the equation

$$2ax-x^2=-2ab-b^2\ ?$$

Ans $\left\{\begin{array}{l} x=\ \ 2a+b. \\ x=-b. \end{array}\right.$

10. What are the values of x in the equation

$$a^2+b^2-2bx+x^2=\frac{m^2x^2}{n^2}\ ?$$

Ans. $\left\{\begin{array}{l} x=\dfrac{n}{n^2-m^2}\left(bn+\sqrt{a^2m^2+b^2m^2-a^2n^2}\right). \\ \\ x=\dfrac{n}{n^2-m^2}\left(bn-\sqrt{a^2m^2+b^2m^2-a^2n^2}\right). \end{array}\right.$

EXAMPLES IN THE THIRD FORM.

1. What are the values of x in the equation

$$x^2+4x=-3\ ?$$

First, by completing the square, we have

$$x^2+4x+4=-3+4=1\ ;$$

and by extracting the square root,

$$x+2=\pm\sqrt{1}=+1 \quad \text{or} \quad -1:$$

hence, $x=-2+1=-1$; or $x=-2-1=-3$

That is, in this form both the roots are negative.

Verification.

If we take the first negative value, the equation

$$x^2+4x=-3$$

gives $$(-1)^2+4(-1)=1-4=-3\ ;$$

and by taking the second value, the equation

$$x^2+4x=-3$$

gives $$(-3)^2+4(-3)=9-12=-3:$$

hence, both values of x satisfy the given equation.

2 What are the values of x in the equation

$$-\frac{x^2}{2}-5x-16=12+\frac{1}{2}x^2+6x.$$

By transposing and reducing, we have

$$-x^2-11x=28\ ;$$

then since the coefficient of the second power of x is negative, we change the signs of all the terms which gives

$$x^2+11x=-28,$$

then by completing the square

$$x^2+11x+30{,}25=2{,}25,$$

hence,

$$x+5{,}5=\pm\sqrt{2{,}25}=+1{,}5 \quad \text{or} \quad -1{,}5.$$

onsequently,

$$x=-4 \quad \text{or} \quad x=-7.$$

3. What are the values of x in the equation

$$-\frac{x^2}{8}-2x-5=\frac{7}{8}x^2+5x+5.$$

Ans. $\begin{cases} x=-2 \\ x=-5. \end{cases}$

4. What are the values of x in the equation

$$2x^2+8x=-2\tfrac{2}{3}-\frac{2}{3}x.$$

Ans. $\begin{cases} x=-4 \\ x=-\tfrac{1}{3}. \end{cases}$

5. What are the values of x in the equation

$$4x^2+\frac{3}{5}x+3x=-14x-3\tfrac{1}{5}-4x^2.$$

Ans. $\begin{cases} x=-2 \\ x=-\tfrac{1}{5}. \end{cases}$

6. What are the values of x in the equation

$$-x^2-4-\frac{3}{4}x=\frac{4x^2}{2}+24x+2.$$

Ans. $\begin{cases} x=-8 \\ x=-\tfrac{1}{4}. \end{cases}$

7. What are the values of x in the equation

$$\frac{1}{9}x^2+7x+20=-\frac{8}{9}x^2-11x-60.$$

Ans. $\begin{cases} x=-10. \\ x=-\ 8. \end{cases}$

8. What are the values of x in the equation

$$\frac{5}{6}x^2-x+\frac{1}{2}=-9\tfrac{1}{8}x-\frac{1}{6}x^2-\frac{1}{2}.$$

Ans. $\begin{cases} x=-8 \\ x=-\frac{1}{8} \end{cases}$

9. What are the values of x in the equation

$$\frac{4}{5}x^2+5x+\frac{1}{4}=-\frac{1}{5}x^2-5\tfrac{1}{10}x-\frac{3}{4}.$$

Ans. $\begin{cases} x=-10 \\ x=-\frac{1}{10}. \end{cases}$

10. What are the values of x in the equation

$$x-x^2-3=6x+1.$$

Ans. $\begin{cases} x=-4 \\ x=-1 \end{cases}$

11. What are the values of x in the equation

$$x^2+4x-90=-93.$$

Ans. $\begin{cases} x=-3 \\ x=-1. \end{cases}$

EXAMPLES IN THE FOURTH FORM.

1. What are the values of x in the equation

$$x^2-8x=-7.$$

By completing the square we have

$$x^2-8x+16=-7+16=9;$$

then by extracting the square root

$$x-4=\pm\sqrt{9}=+3 \quad \text{or} \quad -3;$$

hence,

$$x=+7 \quad \text{or} \quad x=+1.$$

That is, in this form, both the roots are positive.

Verification.

If we take the largest root, the equation

$$x^2-8x=-7 \quad \text{gives} \quad 7^2-8\times7=49-56=-7;$$

and for the smaller, the equation

$$x^2-8x=-7 \quad \text{gives} \quad 1^2-8\times1=1-8=-7:$$

hence, both of the roots will satisfy the equation.

2. What are the values of x in the equation

$$-1\tfrac{1}{2}x^2+3x-10=1\tfrac{1}{2}x^2-18x+\frac{40}{2}.$$

By clearing the fractions, we have

$$-3x^2+6x-20=3x^2-36x+40;$$

then by collecting the like terms

$$-6x^2+42x=60;$$

then by dividing by the coefficient of x^2, and at the same time changing the signs of all the terms, we have

$$x^2-7x=-10.$$

By completing the square, we have

$$x^2-7x+12,25=2,25,$$

and by extracting the square root of both members,

$$x-3,5=\pm\sqrt{2,25}=+1,5 \quad \text{or} \quad -1,5.$$

hence,

$$x=3,5+1,5=5, \quad \text{or} \quad x=3,5-1,5=2.$$

Verification.

If we take the larger root, the equation

$$x^2-7x=-10 \quad \text{gives} \quad 5^2-7\times5=25-35=-10\ ;$$

and if we take the smaller root, the equation

$$x^2-7x=-10 \quad \text{gives} \quad 2^2-7\times2=4-14=-10.$$

3. What are the values of x in the equation

$$-3x+2x^2+1=17\tfrac{4}{5}x-2x^2-3.$$

By transposing and collecting the terms, we have

$$4x^2-20\tfrac{4}{5}x=-4\ ;$$

then dividing by the coefficient of x^2 we have

$$x^2-5\tfrac{1}{5}x=-1.$$

By completing the square, we obtain

$$x^2-5\tfrac{1}{5}x+\frac{169}{25}=-1+\frac{169}{25}=\frac{144}{25},$$

and by extracting the root

$$x^2-2\tfrac{3}{5}=\pm\sqrt{\frac{144}{25}}=+\frac{12}{5} \quad \text{or} \quad -\frac{12}{5}\ ;$$

hence,

$$x=2\tfrac{3}{5}+\frac{12}{5}=5\ ; \quad \text{or,} \quad x=2\tfrac{3}{5}-\frac{12}{5}=\frac{1}{5}.$$

Verification.

If we take the larger root, the equation

$$x^2-5\tfrac{1}{5}x=-1 \quad \text{gives} \quad 5^2-5\tfrac{1}{5}\times5=25-26=-1\ .$$

and if we take the smaller root, the equation

$$x^2-5\tfrac{1}{5}x=-1 \quad \text{gives} \quad \left(\frac{1}{5}\right)^2-5\tfrac{1}{5}\times\frac{1}{5}=\frac{1}{25}-\frac{26}{25}=-1$$

4. What are the values of x in the equation

$$\frac{1}{7}x^2-3x+\frac{1}{2}=-\frac{6}{7}x^2+\frac{1}{4}x-\frac{1}{4}\,?$$

Ans. $\begin{cases} x=3. \\ x=\frac{1}{4}. \end{cases}$

5. What are the values of x in the equation

$$-4x^2-\frac{1}{7}x+1\tfrac{1}{7}=-5x^2+8x\,?$$

Ans. $\begin{cases} x=8. \\ x=\frac{1}{7}. \end{cases}$

6. What are the values of x in the equation

$$-4x^2+\frac{8}{20}x-\frac{1}{40}=-3x^2-\frac{1}{20}x+\frac{1}{40}\,?$$

Ans. $\begin{cases} x=\frac{1}{4} \\ x=\frac{1}{5} \end{cases}$

7. What are the values of x in the equation

$$x^2-10\tfrac{1}{10}x=-1\,?$$

Ans. $\begin{cases} x=10 \\ x=\frac{1}{10}. \end{cases}$

8. What are the values of x in the equation

$$-27x+\frac{17x^2}{5}+100=\frac{2x^2}{5}+12x-26\,?$$

Ans. $\begin{cases} x=7 \\ x=6. \end{cases}$

9. What are the values of x in the equation

$$\frac{8x^2}{3}-22x+15=-\frac{7x^2}{3}+28x-30$$

Ans. $\begin{cases} x=9. \\ x=1. \end{cases}$

10. What are the values of x in the equation

$$2x^2-30x+3=\;-x^2+3\tfrac{3}{10}x-\frac{3}{10}\,?$$

Ans. $\begin{cases} x=11 \\ x= \end{cases}$ [illegible]

Properties of the Roots.

128. We have thus far, only explained the methods of finding the roots of an equation of the second degree. We are now going to show some of the properties of these roots

The first form.

129. The first form

$$x^2+2px=q$$

gives 1st root $\quad x=-p+\sqrt{q+p^2},$

2nd root $\quad x=-p-\sqrt{q+p^2},$

and their sum $\quad =-2p.$

Since, in this form q is supposed positive, the quantity $q+p^2$ under the radical sign will be greater than p^2, and hence its root will be greater than p. Consequently the first root, which is equal to the difference between p and the radical, will be positive and less than $\sqrt{q+p^2}$. In the second root, p and the radical have the same sign; hence, the second root will be equal to their sum and negative. If we multiply the two roots together, we have

$$\begin{array}{l} -p+\sqrt{q+p^2} \\ -p-\sqrt{q+p^2} \\ \hline +p^2-p\sqrt{q+p^2} \\ \qquad +p\sqrt{q+p^2}-q-p^2 \end{array}$$

Product equal to . . . . . . $-q.$

QUEST.—**129.** In the first form, have the roots the same or contrary signs? What is the sign of the first root? What of the second? Which is the greater? What is their sum equal to? What is their product equal to?

Hence we conclude,

1st. *That in the first form one of the roots is always positive, and the other negative.*

2nd. *That the positive root is numerically less than the negative.*

3rd. *That the sum of the two roots is equal to the coefficient of* x *in the second term, taken with a contrary sign.*

4th. *That the product of the two roots is equal to the known term in the second member, taken with a contrary sign.*

EXAMPLES.

1. In the equation

$$x^2+x=20,$$

we find the roots to be 4 and −5. Their sum is −1, and their product −20.

2. In the equation

$$x^2+2x=3,$$

we find the roots to be 1 and −3. Their sum is equal to −2, and their product to −3.

3. The roots of the equation

$$x^2+x=90,$$

are +9 and −10. Their sum is −1, and their product −90.

4. The roots of the equation

$$x^2+4x=60,$$

are 6 and −10. Their sum is −4, and their product is −60.

Let these principles be applied to each of the examples under "EXAMPLES IN THE FIRST FORM."

Second Form.

130. The second form is,

$$x^2-2px=q\,;$$

and by resolving the equation we find

1st root, $\qquad x=+p+\sqrt{q+p^2}$

2nd root, $\qquad x=+p-\sqrt{q+p^2}$

and their sum $\qquad = 2p.$

In this form, the first root is positive and the second negative. If we multiply the two roots together, we have

$$(p+\sqrt{q+p^2})\times(p-\sqrt{q+p^2})=-q.$$

Hence we conclude,

1st. *That in the second form one of the roots is positive and the other negative.*

2nd. *That the positive root is numerically greater than the negative.*

3rd. *That the sum of the roots is equal to the coefficient of* x *in the second term, taken with a contrary sign.*

4th. *That the product of the roots is equal to the known term in the second member, taken with a contrary sign.*

QUEST.—**130**. What is the sign of the first root in the second form? What is the sign of the second? Which is the greater? What is their sum equal to? What is their product equal to?

EXAMPLES.

1. The roots of the equation

$$x^2-x=12,$$

are $+4$ and -3. Their sum is $+1$, and their product -12.

2. The roots of the equation

$$x^2-9\tfrac{9}{10}x=1,$$

are $+10$ and $-\frac{1}{10}$. Their sum is $9\frac{9}{10}$, and their product is -1.

3. The roots of the equation

$$x^2-6x=16,$$

are $+8$ and -2. Their sum is $+6$, and their product is -16.

4. The roots of the equation

$$x^2-11x=80,$$

are $+16$ and -5. Their sum is $+11$, and their product is -80.

Let these principles be applied to each of the examples under "EXAMPLES IN THE SECOND FORM."

Third Form.

131. The third form is,

$$x^2+2px=-q;$$

and by resolving the equation we find,

1st root, $x=-p+\sqrt{-q+p^2},$

2nd root, $x=-p-\sqrt{-q+p^2}$

Their sum is $=-2p$

In this form, the quantity under the radical being less than p^2, its root will be less than p: hence both the roots will be negative, and the first will be numerically the least.

If we multiply the roots together, we have

$$(-p+\sqrt{-q+p^2})\times(-p-\sqrt{-q+p^2})=+q.$$

Hence we conclude,

1st. *That in the third form both the roots are negative.*

2nd. *That the first root is numerically less than the second.*

3rd. *That the sum of the two roots is equal to the coefficient of* x *in the second term, taken with a contrary sign.*

4th. *That the product of the roots is equal to the known term in the second member, taken with a contrary sign.*

EXAMPLES.

1. The roots of the equation

$$x^2+9x=-20,$$

are -4 and -5. Their sum is -9, and their product $+20$.

2. The roots of the equation

$$x^2+13x=-42,$$

are -6 and -7. Their sum is -13, and their product $+42$.

QUEST.—**131**. In the third form, what are the signs of the roots? Which root is the least? What is the sum of the roots equal to? What is their product equal to?

3. The roots of the equation

$$x^2+2\frac{3}{4}x=-1\frac{1}{2},$$

are $-\frac{3}{4}$ and -2. Their sum is $-2\frac{3}{4}$, and their product $+1\frac{1}{2}$.

4. The roots of the equation

$$x^2+5x=-6,$$

are -2 and -3. Their sum is -5, and their product is $+6$

Let these principles be applied to each of the examples under "EXAMPLES IN THE THIRD FORM."

Fourth Form.

132. The fourth form is,

$$x^2-2px=-q;$$

and by resolving the equation we find,

1st root, $\quad x=p+\sqrt{-q+p^2}$

2nd root, $\quad x=p-\sqrt{-q+p^2}$

Their sum is $\quad =2p.$

In this form, as well as in the third, the quantity under the radical being less than p^2, its root will be less than p: hence both the roots will be positive, and the first will be the greatest.

If we multiply the two roots together, we have

$$(p+\sqrt{-q+p^2})\times(p-\sqrt{-q+p^2})=+q.$$

Hence we conclude,

1st. *That in the fourth form both the roots are positive.*

2nd. *That the first root is greater than the second.*

3rd. *That the sum of the roots is equal to the coefficient of* x *in the second term, taken with a contrary sign.*

4th. *That the product of the roots is equal to the known term in the second member, taken with a contrary sign.*

EXAMPLES.

1. The roots of the equation

$$x^2-7x=-12,$$

are $+4$ and $+3$. Their sum is $+7$ and their product $+12$.

2. The roots of the equation

$$x^2-14x=-24,$$

are $+12$ and $+2$. Their sum is $+14$ and their product $+24$.

3. The roots of the equation

$$x^2-20x=-36,$$

are $+18$ and $+2$. Their sum is $+20$ and their product $+36$.

4. The roots of the equation

$$x^2-17x=-42,$$

are $+14$ and $+3$. Their sum is $+17$ and their product $+42$.

QUEST.—**132** In the fourth form, what are the signs of the roots? Which root is the greatest? What is the sum of the roots equal to? What is their product equal to?

133. In the third and fourth forms the values of x sometimes become imaginary, and in such cases it is necessary to know how the results are to be interpreted.

If we have $q>p^2$, *that is, if the known term is greater than half the coefficient of* x *squared*, it is plain that $\sqrt{-q+p^2}$ will be imaginary, since the quantity under the radical will be negative. Under this supposition the values of x in the third and fourth forms will be imaginary.

We will now show that, when in the third and fourth forms, we have $q>p^2$, the conditions of the question will be incompatible with each other.

134. Before showing this it will be necessary to establish a proposition on which it depends: viz.

If a given number be decomposed into two parts and those parts multiplied together, the product will be the greatest possible when the parts are equal.

Let $2p$ be the number to be decomposed, and d the difference of the parts. Then

$p+\frac{d}{2}=$ the greater part (page 80, Ex. 7.)

and $p-\frac{d}{2}=$ the less part.

and $p^2-\frac{d^2}{4}=\text{P}$, their product (Art. **40**.)

Now it is plain that P will increase as d diminishes, and that it will be the greatest possible when $d=0$: that is.

$$p\times p=p^2 \text{ is the greatest product.}$$

QUEST.—**133**. In which forms do the values of x become imaginary? When will the values of x be imaginary? Why will the values of x be then imaginary?

Now, since in the equation

$$x^2-2px=-q$$

$2p$ is the sum of the roots, and q their product, it follows that q can never be greater than p^2. The conditions of the equation, therefore, fix a limit to the value of q, and if we make $q>p^2$, we express by the equation a condition which cannot be fulfilled, and, this contradiction is made apparent by the values of x becoming imaginary. Hence we may conclude that,

When the values of the unknown quantity are imaginary, the conditions of the question are incompatible with each other.

EXAMPLES.

1. Find two numbers whose sum shall be 12 and product 46.

Let x and y be the numbers.

By the 1st condition, $x+y=12$;

and by the 2d, $xy=46$.

The first equation gives

$$x=12-y.$$

Substituting this value for x in the second, we have

$$12y-y^2=46;$$

and changing the signs of the terms, we have

$$y^2-12y=-46.$$

QUEST.—**134**. What is the proposition demonstrated in Article **134**? If the conditions of the question are incompatible, how will the values of the unknown quantity be?

Then by completing the square

$$y^2-12y+36=-46+36=-10$$

which gives $y=6+\sqrt{-10}$,

and $y=6-\sqrt{-10}$;

both of which values are imaginary, as indeed they should be, since the conditions are incompatible.

2. The sum of two numbers is 8, and their product 20: what are the numbers?

Denote the numbers by x and y.

By the first condition,

$$x+y=8;$$

and by the second, $xy=20$.

The first equation gives

$$x=8-y$$

Substituting this value of x in the second, we have

$$8y-y^2=20;$$

changing the signs, and completing the square, we have

$$y^2-8y+16=-4;$$

and by extracting the root,

$$y=4+\sqrt{-4} \text{ and } y=4-\sqrt{-4}.$$

These values of y may be put under the forms (Art. **106**)

$$y=4+2\sqrt{-1} \text{ and } y=4-2\sqrt{-1}.$$

3. What are the values of x in the equation

$$x^2+2x=-10.$$

Ans. $\begin{cases} x=-1+3\sqrt{-1} \\ x=-1-3\sqrt{-1} \end{cases}$

Examples with more than one unknown quantity.

1. Given $\begin{cases} x+y=14 \\ x^2+y^2=100 \end{cases}$ to find x and y.

By transposing y in the first equation, we have

$$x=14-y;$$

and by squaring both members,

$$x^2=196-28y+y^2.$$

Substituting this value for x^2 in the 2nd equation, we have

$$196-28y+y^2+y^2=100;$$

from which we have

$$y^2-14y=-48;$$

and by completing the square,

$$y^2-14y+49=1;$$

and by extracting the square root,

$$y-7=\pm\sqrt{1}=+1 \text{ or } -1:$$

hence, $\quad y=7+1=8, \text{ or } y=7-1=6.$

If we take the larger value, we find $x=6$; and if we take the smaller, we find $x=8$.

Verification.

For the largest value, $y=8$, the equation

$$x+y=14 \quad \text{gives} \quad 6+8=14;$$

and $\quad x^2+y^2=100 \quad \text{gives} \quad 36+64=100.$

For the value $y=6$, the equation

$$x+y=14 \quad \text{gives} \quad 8+6=14;$$

and $\quad x^2+y^2=100 \quad \text{gives} \quad 64+36=100.$

Hence, both sets of values will satisfy the given equation

2. Given $\left\{\begin{array}{l} x-y=3 \\ x^2-y^2=45 \end{array}\right\}$ to find x and y.

Transposing y in the first equation, we have

$$x=3+y\ ;$$

and then squaring both members,

$$x^2=9+6y+y^2.$$

Substituting this value for x^2 in the second equation, we have

$$9+6y+y^2-y^2=45\ ;$$

whence we have

$$6y=36 \quad \text{and} \quad y=6.$$

Substituting this value of y in the first equation, we have

$$x-6=3,$$

and consequently $x=3+6=9$.

Verification.

$x-y=3$ gives $9-6=3$;

and $x^2-y^2=45$ gives $81-36=45$.

3. Given $\left\{\begin{array}{l} x^2+3xy=22 \\ x^2+3xy+2y^2=40 \end{array}\right\}$ to find x and y.

Subtracting the first equation from the second, we have

$$2y^2=18,$$

which gives $y^2=9$,

and $y=+3$ or -3.

Substituting the plus value in the first equation, we have

$$x^2+9x=22\ ;$$

from which we find

$$x=+2 \quad \text{and} \quad x=-11.$$

If we take the negative value, $y=-3$, we have from the first equation,

$$x^2-9x=22;$$

from which we find

$$x=+11 \quad \text{and} \quad x=-2.$$

Verification.

For the values $y=+3$ and $x=+2$, the equation

$$x^2+3xy=22$$

gives $$2^2+3\times2\times3=4+18=22:$$

and for the second value, $x=-11$, the same equation

$$x^2+3xy=22$$

gives $$(-11)^2+3\times-11\times3=121-99=22.$$

If now we take the second value of y, that is, $y=-3$ and the corresponding values of x, viz, $x=+11$, and $x=-2$; for $x=+11$, the equation

$$x^2+3xy=22$$

gives $$11^2+3\times11\times-3=121-99=22;$$

and for $x=-2$, the same equation

$$x^2+3xy=22$$

gives $$(-2)^2+3\times-2\times-3=4+18=22.$$

4. Given $\left\{\begin{array}{lr} xz=y^2 & (1) \\ x+y+z=7 & (2) \\ x^2+y^2+z^2=21 & (3) \end{array}\right\}$ to find x, y, and z.

Transposing y in the second equation, we have

$$x+z=7-y \quad (4);$$

then squaring the members, we have

$$x^2+2xz+z^2=49-14y+y^2.$$

If now we substitute for $2xz$ its value taken from the first equation, we have

$$x^2+2y^2+z^2=49-14y+y^2;$$

and cancelling y^2 in each member, there results

$$x^2+y^2+z^2=49-14y.$$

But, from the third equation we see that each member of the last equation is equal to 21 : hence

$$49-14y=21,$$

and $$14y=49-21=28.$$

hence, $$y=\frac{28}{14}=2.$$

Placing this value for y in equation (1) gives

$$xz=4;$$

and placing it in equation (4) gives

$$x+z=5, \quad \text{and} \quad x=5-z.$$

Substituting this value of x in the previous equation, we obtain

$$5z-z^2=4 \quad \text{or} \quad z^2-5z=-4;$$

and by completing the square, we have

$$z^2-5z+6,25=2,5,$$

and $$z-2,5=\pm\sqrt{2,5}=+1,5 \quad \text{or} \quad -1,5;$$

hence, $$z=2,5+1,5=4 \quad \text{or} \quad z=+2,5-1,5=1.$$

If we take the value

$$z=4, \quad \text{we find} \quad x=1:$$

if we take the less value

$$z=1, \quad \text{we find} \quad x=4.$$

3. Given $x+\sqrt{xy}+y=19$ and $x^2+xy+y^2=133$ } to find x and y.

Dividing the second equation by the first, we have

$$x-\sqrt{xy}+y=7$$

but $x+\sqrt{xy}+y=19$

hence, by addition, $2x+2y=26$

or $x+y=13$

and substituting in 1st equa. $\sqrt{xy}+13=19$

or $\sqrt{xy}=6$

and by squaring $xy=36$

From 2d equation, $x^2+xy+y^2=133$

and from the last $3xy=108$

Subtracting $x^2-2xy+y^2=25$

hence, $x-y=\pm 5$

but $x+y=13$

hence $x=9$ or 4; and $y=4$ or 9.

6. Given the sum of two numbers equal to a, and the sum of their cubes equal to c, to find the numbers

By the conditions $\begin{cases} x+y=a \\ x^3+y^3=c \end{cases}$

Putting $x=s+z$, and $y=s-z$, we have

$$a=2s, \quad \text{or} \quad s=\frac{a}{2};$$

and
$$\begin{cases} x^3=s^3+3s^2z+3sz^2+z^3 \\ y^3=s^3-3s^2z+3sz^2-z^3. \end{cases}$$

hence, by addition, $x^3+y^3=2s^3+6sz^2=c$,

whence $$z^2=\frac{c-2s^3}{6s} \quad \text{and} \quad z=\pm\sqrt{\frac{c-2s^3}{6s}};$$

or $$x=s\pm\sqrt{\frac{c-2s^3}{6s}}; \quad \text{and} \quad y=s\mp\sqrt{\frac{c-2s^3}{6s}};$$

or by putting for s its value,

$$x=\frac{a}{2}\pm\sqrt{\left(\frac{c-\frac{a^3}{4}}{3a}\right)}=\frac{a}{2}\pm\sqrt{\frac{4c-a^3}{12a}},$$

and $$y=\frac{a}{2}\mp\sqrt{\left(\frac{c-\frac{a^3}{4}}{3a}\right)}=\frac{a}{2}\mp\sqrt{\frac{4c-a^3}{12a}}.$$

NOTE.—What are the numbers when $a=5$ and $c=35$. What are the numbers when $a=9$ and $c=243$.

QUESTIONS.

1. Find a number such, that twice its square, added to three times the number, shall give 65.

Let x denote the unknown number. Then the equation of the problem will be

$$2x^2+3x=65,$$

whence

$$x=-\frac{3}{4}\pm\sqrt{\frac{65}{2}+\frac{9}{16}}=-\frac{3}{4}\pm\frac{23}{4}.$$

Therefore,

$$x=-\frac{3}{4}+\frac{23}{4}=5, \text{ and } x=-\frac{3}{4}-\frac{23}{4}=-\frac{13}{2}.$$

Both these values satisfy the question in its algebraic sense For,

$$2\times(5)^2+3\times 5=2\times 25+15=65\ ;$$

and $$2\left(-\frac{13}{2}\right)^2+3\times -\frac{13}{2}=\frac{169}{2}-\frac{39}{2}=\frac{130}{2}=65.$$

REMARK.—If we wish to restrict the enunciation to its arithmetical sense, we will first observe, that when x is replaced by $-x$, in the equation $2x^2+3x=65$, the sign of the second term $3x$ only, is changed, because $(-x)^2=x^2$.

Therefore, instead of obtaining $x=-\frac{3}{4}\pm\frac{23}{4}$, we should find $x=\frac{3}{4}\pm\frac{23}{4}$,. or $x=\frac{13}{2}$, and $x=-5$, values which only differ from the preceding by their signs. Hence, we may say that the negative solution $-\frac{13}{2}$, considered independently of its sign, satisfies this new enunciation, viz: *To find a number such, that twice its square, diminished by three times the number, shall give* 65. In fact, we have

$$2\times\left(\frac{13}{2}\right)^2-3\times\frac{13}{2}=\frac{169}{2}-\frac{39}{2}=65.$$

REMARK.—The root which results from giving the plus sign to the radical, generally resolves the question both in its arithmetical and algebraic sense, while the second root resolves it in its algebraic sense only.

Thus, in the example, it was required to find a number of which twice the square *added* to three times the number shall give 65. Now, in the arithmetical sense, *added* means increased; but in the algebraic sense it implies diminution, when the quantity added is negative. In this sense, the second root satisfies the enunciation.

2. A certain person purchased a number of yards of cloth for 240 cents. If he had received 3 yards less of the same cloth for the same sum, it would have cost him 4 cents more per yard. How many yards did he purchase?

Let $x=$ the number of yards purchased.

Then $\frac{240}{x}$ will express the price per yard.

If, for 240 cents, he had received 3 yards less, that is $x-3$ yards, the price per yard, under this hypothesis, would have been represented by $\frac{240}{x-3}$. But, by the enunciation, this last cost would exceed the first by 4 cents. Therefore, we have the equation

$$\frac{240}{x-3}-\frac{240}{x}=4;$$

whence, by reducing $x^2-3x=180$,

$$x=\frac{3}{2}\pm\sqrt{\frac{9}{4}+180}=\frac{3\pm27}{2};$$

therefore $x=15$ and $x=-12$.

The value $x=15$ satisfies the enunciation; for, 15 yards for 240 cents gives $\frac{240}{15}$, or 16 cents for the price of one yard, and 12 yards for 240 cents, gives 20 cents for the price of one yard, which exceeds 16 by 4.

As to the second solution, we can form a new enunciation, with which it will agree. For, going back to the equation, and changing x into $-x$, it becomes

$$\frac{240}{-x-3}-\frac{240}{-x}=4, \quad \text{or} \quad \frac{240}{x}-\frac{240}{x+3}=4,$$

an equation which may be considered the algebraic translation of this problem, viz: *A certain person purchased a number of yards of cloth for* 240 *cents: if he had paid the same sum for* 3 *yards more, it would have cost him* 4 *cents less per yard. How many yards did he purchase?*

Ans. $x=12$, and $x=-15$.

3. A man bought a horse, which he sold after some time for 24 dollars. At this sale, he loses as much per cent. upon the price of his purchase as the horse cost him. What did he pay for the horse?

Let x denote the number of dollars that he paid for the horse, $x-24$ will express the loss he sustained. But as he lost x per cent. by the sale, he must have lost $\frac{x}{100}$ upon each dollar, and upon x dollars he loses a sum denoted by $\frac{x^2}{100}$; we have then the equation

$$\frac{x^2}{100}=x-24, \quad \text{whence} \quad x^2-100x=-2400.$$

and $$x=50\pm\sqrt{2500-2400}=50\pm10.$$

Therefore, $$x=60 \text{ and } x=40.$$

Both of these values satisfy the question.

For, in the first place, suppose the man gave $60 for the horse and sold him for 24, he loses 36. Again, from the enunciation, he should lose 60 *per cent.* of 60, that is,

$\frac{60}{100}$ of 60, or $\frac{60\times 60}{100}$, which reduces to 36 ; therefore, 60 satisfies the enunciation.

Had he paid $40, he would have lost $16 by the sale ; for, he should lose 40 *per cent.* of 40, or $40\times\frac{40}{100}$, which reduces to 16 ; therefore, 40 verifies the enunciation.

4. A man being asked his age, said the square root of my own age is half the age of my son, and the sum of our ages is 80 years : what was the age of each ?

Let $x=$ the age of the father.
$y=$ that of the son.

Then by the first condition

$$\sqrt{x}=\frac{y}{2},$$

and by the second condition

$$x+y=80.$$

If we take the first equation

$$\sqrt{x}=\frac{y}{2},$$

and square both members, we have

$$x=\frac{y^2}{4}.$$

If we transpose y in the second, we have

$$x=80-y:$$

from which we find

$$y=-2\pm\sqrt{324}=16;$$

by taking the plus root, which answers to the question in its arithmetical sense. Substituting this value, we find $x=64$.

Ans. { Father's age 64
Son's 16

5. Find two numbers, such that the sum of their products by the respective numbers a and b, may be equal to $2s$, and that their product may be equal to p.

Let x and y be the required numbers, we have the equations

$$ax+by=2s.$$

and

$$xy=p.$$

From the first $$y=\frac{2s-ax}{b};$$

whence, by substituting in the second, and reducing,

$$ax^2-2sx=-bp.$$

Therefore, $$x=\frac{s}{a}\pm\frac{1}{a}\sqrt{s^2-abp},$$

and consequently,

$$y=\frac{s}{b}\mp\frac{1}{b}\sqrt{s^2-abp}.$$

This problem is susceptible of two direct solutions, because s is evidently $>\sqrt{s^2-abp}$; but in order that they may be real, it is necessary that $s^2>$ or $=abp$.

Let $a=b=1$; the values of x and y reduce to

$$x=s\pm\sqrt{s^2-p} \quad \text{and} \quad y=s\mp\sqrt{s^2-p}.$$

Whence we see, that the two values of x are equal to those of y, taken in an inverse order; which shows, that if $s+\sqrt{s^2-p}$ represents the value of x, $s-\sqrt{s^2-p}$ will represent the corresponding value of y, and reciprocally.

This circumstance is accounted for, by observing that in this particular case the equations reduce to

$$\left\{\begin{array}{r} x+y=2s, \\ xy=p; \end{array}\right\}$$

and then the question is reduced to, *finding two numbers of which the sum is $2s$, and their product p*, or in other words *to divide a number $2s$, into two such parts, that their product may be equal to a given number p.*

Let us now suppose

$$2s=14 \quad \text{and} \quad p=48:$$

what will then be the values of x and y ?

Ans. $\begin{cases} x=8 \text{ or } 6 \\ y=6 \text{ or } 8 \end{cases}$

6. A grazier bought as many sheep as cost him £60, and after reserving fifteen out of the number, he sold the remainder for £54, and gained 2*s*. a head on those he sold: how many did he buy? *Ans.* 75.

7. A merchant bought cloth for which he paid £33 15*s*., which he sold again at £2 8*s*. per piece, and gained by the bargain as much as one piece cost him: how many pieces did he buy? *Ans.* 15.

8. What number is that, which, being divided by the product of its digits, the quotient is 3; and if 18 be added to it, the digits will be inverted? *Ans.* 24.

9. To find a number, such that if you subtract it from 10, and multiply the remainder by the number itself, the product shall be 21. *Ans.* 7 or 3.

10. Two persons, A and B, departed from different places at the same time, and travelled towards each other. On meeting, it appeared that A had travelled 18 miles more than B; and that A could have gone B's journey in $15\frac{3}{4}$ days, but B would have been 28 days in performing A's journey How far did each travel?

Ans. $\begin{cases} \text{A } 72 \text{ miles} \\ \text{B } 54 \text{ miles} \end{cases}$

11. There are two numbers whose difference is 15, and half their product is equal to the cube of the lesser number. What are those numbers? *Ans.* 3 and 18.

12. What two numbers are those whose sum, multiplied by the greater, is equal to 77; and whose difference, multiplied by the lesser, is equal to 12?

Ans. 4 and 7, or $\frac{3}{2}\sqrt{2}$ and $\frac{11}{2}\sqrt{2}$.

13. To divide 100 into two such parts, that the sum of their square roots may be 14. *Ans.* 64 and 36.

14. It is required to divide the number 24 into two such parts, that their product may be equal to 35 times their difference. *Ans.* 10 and 14.

15. The sum of two numbers is 8, and the sum of their cubes 152. What are the numbers? *Ans.* 3 and 5.

16. Two merchants each sold the same kind of stuff; the second sold 3 yards more of it than the first, and together they receive 35 dollars. The first said to the second, "I would have received 24 dollars for your stuff;" the other replied, "And I should have received $12\frac{1}{2}$ dollars for yours." How many yards did each of them sell?

Ans. $\left\{\begin{array}{l}\text{1st merchant } x=15 \\ \text{2nd } \quad ,, \quad y=18\end{array}\right.$ or $\begin{array}{l}x=5. \\ y=8.\end{array}$

17. A widow possessed 13,000 dollars, which she divided into two parts, and placed them at interest, in such a manner, that the incomes from them were equal. If she had put out the first portion at the same rate as the second, she would have drawn for this part 360 dollars interest; and if she had placed the second out at the same rate as the first, she would have drawn for it 490 dollars interest. What were the two rates of interest?

Ans. 7 and 6 per cent.

CHAPTER VII.

Of Proportions and Progressions.

135. Two quantities of the same kind may be compared together in two ways :—

1st. By considering *how much* one is greater or less than the other, which is shown by their difference ; and,

2nd. By considering *how many times* one is greater or less than the other, which is shown by their quotient.

Thus, in comparing the numbers 3 and 12 together with respect to their difference, we find that 12 *exceeds* 3 by 9 ; and in comparing them together with respect to their quotient, we find that 12 *contains* 3 four times, or that 12 is 4 times as great as 3.

The first of these methods of comparison is called *Arithmetical Proportion*, and the second *Geometrical Proportion.*

Hence, *Arithmetical Proportion considers the relation of quantities with respect to their difference, and Geometrical Proportion the relation of quantities with respect to their quotient.*

Quest.—**135**. In how many ways may two quantities be compared together? What does the first method consider? What the second? What is the first of these methods called? What is the second called? How then do you define the two proportions?

Of Arithmetical Proportion and Progression.

136. If we have four numbers, 2, 4, 8, and 10, of which the difference between the first and second is equal to the difference between the third and fourth, these numbers are said to be in arithmetical proportion. The first term 2 is called an *antecedent*, and the second term 4, with which it is compared, a *consequent.* The number 8 is also called an antecedent, and the number 10, with which it is compared, a consequent.

When the difference between the first and second is equal to the difference between the third and fourth, the four numbers are said to be in proportion. Thus, the numbers

2, 4, 8, 10,

are in arithmetical proportion.

137. When the difference between the first antecedent and consequent is the same as between any two adjacent terms of the proportion, the proportion is called an *arithmetical progression.* Hence, a *progression by differences*, or an *arithmetical progression*, is a series in which the successive terms continually increase or decrease by a constant number, which is called the *common difference* of the progression.

Thus, in the two series

1, 4, 7, 10, 13, 16, 19, 22, 25, . . .

60, 56, 52, 48, 44, 40, 36, 32, 28, . . .

QUEST.—**136.** When are four numbers in arithmetical proportion? What is the first called? What is the second called? What is the third called? What is the fourth called?

the first is called an *increasing progression*, of which the common difference is 3, and the second a *decreasing progression*, of which the common difference is 4.

In general, let a, b, c, d, e, f, . . . designate the terms of a progression by differences; it has been agreed to write them thus:

$$a \;.\; b \;.\; c \;.\; d \;.\; e \;.\; f \;.\; g \;.\; h \;.\; i \;.\; k \;.\;.\;.$$

This series is read, a is to b, as b is to c, as c is to d, as d is to e, &c. This is a series of *continued equi-differences*, in which each term is at the same time a consequent and antecedent, with the exception of the first term, which is only an *antecedent*, and the last, which is only a *consequent*.

138. Let r represent the common difference of the progression

$$a \;.\; b \;.\; c \;.\; d \;.\; e \;.\; f \;.\; g \;.\; h, \text{ \&c},$$

which we will consider increasing.

From the definition of the progression, it evidently follows that

$$b=a+r, \quad c=b+r=a+2r, \quad d=c+r=a+3r;$$

and, in general, any term of the series is equal to *the first term plus as many times the common difference as there are preceding terms.*

Thus, let l be any term, and n the number which marks the place of it: the expression for this *general term* is

$$l=a+(n-1)r.$$

QUEST.—**137**. What is an arithmetical progression? What is the number called by which the terms are increased or diminished? What is an increasing progression? What is a decreasing progression? Which term is only an antecedent? Which only a consequent?

Hence, for finding the last term, we have the following

RULE.

I. *Multiply the common difference by one less than the number of terms.*

II. *To the product add the first term: the sum will be the last term.*

EXAMPLES.

The formula $l=a+(n-1)r$ serves to find any term whatever, without our being obliged to determine all those which precede it.

1. If we make $n=1$, we have $l=a$; that is, the series will have but one term.

2. If we make $n=2$, we have $l=a+r$; that is, the series will have two terms, and the second term is equal to the first plus the common difference.

3. If $a=3$ and $r=2$, what is the 3rd term? *Ans.* 7.

4. If $a=5$ and $r=4$, what is the 6th term? *Ans.* 25.

5. If $a=7$ and $r=5$, what is the 9th term? *Ans.* 47.

6. If $a=8$ and $r=5$, what is the 10th term? *Ans.* 53.

7. If $a=20$ and $r=4$, what is the 12th term? *Ans.* 64.

8. If $a=40$ and $r=20$, what is the 50th term? *Ans.* 1020.

QUEST.—**138.** Give the rule for finding the last term of a series when the progression is increasing.

9. If $a=45$ and $r=30$, what is the 40th term?

Ans. 1215

10. If $a=30$ and $r=20$, what is the 60th term?

Ans. 1210

11. If $a=50$ and $r=10$, what is the 100th term?

Ans. 1040

12. To find the 50th term of the progression

$$1 . 4 . 7 . 10 . 13 . 16 . 19 \ldots,$$

we have $$l=1+49\times3=148.$$

13. To find the 60th term of the progression

$$1 . 5 . 9 . 13 . 17 . 21 . 25 \ldots,$$

we have $$l=1+59\times4=237.$$

139. If the progression were a decreasing one, we should have

$$l=a-(n-1)r.$$

Hence, to find the last term of a decreasing progression, we have the following

RULE.

I. *Multiply the common difference by one less than the number of terms.*

II. *Subtract the product from the first term. the remainder will be the last term*

QUEST.—**139.** Give the rule for finding the last term of a series when the progression is decreasing.

EXAMPLES.

1. The first term of a decreasing progression is 60, the number of terms 20, and the common difference 3 : what is the last term ?

$l=a-(n-1)r$ gives $l=60-(20-1)3=60-57=3$.

2. The first term is 90, the common difference 4, and the number of terms 15 : what is the last term ? *Ans.* 34.

3. The first term is 100, the number of terms 40, and the common difference 2 : what is the last term ? *Ans.* 22.

4. The first term is 80, the number of terms 10, and the common difference 4 : what is the last term ? *Ans.* 44.

5. The first term is 600, the number of terms 100, and the common difference 5 : what is the last term ?

Ans. 105.

6. The first term is 800, the number of terms 200, and the common difference 2 : what is the last term ?

Ans. 402.

140. A progression by differences being given, it is proposed to prove that, *the sum of any two terms, taken at equal distances from the two extremes, is equal to the sum of the two extremes.*

That is, if we have the progression

2 . 4 . 6 . 8 . 10 . 12,

we wish to prove that

$4+10$ or $6+8$

is equal to the sum of the two extremes 2 and 12.

Let $a \,.\, b \,.\, c \,.\, d \,.\, e \,.\, f \,.\, .\, .\, .\, i \,.\, k \,.\, l$ be the proposed progression, and n the number of terms.

We will first observe that, if x denotes a term which has p terms before it, and y a term which has p terms after it we have, from what has been said,

$$x=a+p\times r,$$

and $$y=l-p\times r\,;$$

whence, by addition, $x+y=a+l$.

Which demonstrates the proposition.

Referring this proof to the previous example, if we suppose, in the first place, x to denote the second term 4, then y will denote the term 10, next to the last. If x denotes the 3rd term 6, then y will denote 8, the third term from the last.

Having proved the first part of the proposition, write the progression below itself, but in an inverse order, viz:

$$a \,.\, b \,.\, c \,.\, d \,.\, e \,.\, f \,.\, .\, .\, i \,.\, k \,.\, l.$$

$$l \,.\, k \,.\, i \,.\, .\, .\, .\, .\, .\, .\, .\, .\, .\, c \,.\, b \,.\, a.$$

Calling S the sum of the terms of the first progression, $2S$ will be the sum of the terms in both progressions, and we shall have

$$2S=(a+l)+(b+k)+(c+i)\ \ldots\ +(i+c)+(k+b)+(l+a).$$

Now, since all the parts $a+l$, $b+k$, $c+i$ . . . are equal to each other, and their number equal to n,

$$2S=(a+l)n, \quad \text{or} \quad S=\left(\frac{a+l}{2}\right)n.$$

Hence, for finding the sum of an arithmetical series, we have the following

RULE.

I. *Add the two extremes together, and take half their sum.*

II. *Multiply the half-sum by the number of terms; the product will be the sum of the series.*

EXAMPLES.

1. The extremes are 2 and 16, and the number of terms 8: what is the sum of the series?

$$S=\left(\frac{a+l}{2}\right)\times n, \quad \text{gives} \quad S=\frac{2+16}{2}\times 8=72.$$

2. The extremes are 3 and 27, and the number of terms 12: what is the sum of the series? *Ans.* 180.

3. The extremes are 4 and 20, and the number of terms 10: what is the sum of the series? *Ans.* 120.

4. The extremes are 100 and 200, and the number of terms 80: what is the sum of the series? *Ans.* 12000.

5. The extremes are 500 and 60, and the number of terms 20: what is the sum of the series? *Ans.* 5600.

6. The extremes are 800 and 1200, and the number of terms 50: what is the sum of the series? *Ans.* 50000.

QUEST.—**140.** In every progression, what is the sum of the two extremes equal to? What is the rule for finding the sum of an arithmetical series?

141. In arithmetical proportion there are five numbers to be considered :—

1st. The first term, a.
2nd. The common difference, r.
3rd. The number of terms, n.
4th. The last term, l.
5th. The sum, S.

The formulas

$$l=a+(n-1)r \quad \text{and} \quad S=\left(\frac{a+l}{2}\right)\times n$$

contain five quantities, a, r, n, l, and S, and consequently give rise to the following general problem, viz : *Any three of these five quantities being given, to determine the other two.*

We already know the value of S in terms of a, n, and r

From the formula

$$l=a+(n-1)r,$$

we find $$a=l-(n-1)r.$$

That is : *The first term of an increasing arithmetical progression is equal to the last term, minus the product of the common difference by the number of terms less one.*

From the same formula, we also find

$$r=\frac{l-a}{n-1}.$$

That is : *In any arithmetical progression, the common difference is equal to the difference between the two extremes divided by the number of terms less one.*

QUEST.—**141**. How many numbers are considered in arithmetical proportion? What are they? In every arithmetical progression, what is the common difference equal to?

The last term is 16, the first term 4, and the number of terms 5 : what is the common difference ?

The formula $$r=\frac{l-a}{n-1}$$

gives $$r=\frac{16-4}{4}=3.$$

2. The last term is 22, the first term 4, and the number of terms 10: what is the common difference ? *Ans.* 2.

142. The last principle affords a solution to the following question :

To find a number m *of arithmetical means between two given numbers* a *and* b.

To resolve this question, it is first necessary to find the common difference. Now, we may regard a as the first term of an arithmetical progression, b as the last term, and the required means as intermediate terms. The number of terms of this progression will be expressed by $m+2$.

Now, by substituting in the above formula, b for l, and $m+2$ for n, it becomes

$$r=\frac{b-a}{m+2-1}=\frac{b-a}{m+1};$$

that is, *the common difference* of the required progression is obtained by dividing the difference between the given numbers a and b, by one more than the required number of means.

QUEST.—**142.** How do you find any number of arithmetical means between two given numbers ?

Having obtained the common difference, form the second term of the progression, or the *first arithmetical mean*, by adding r, or $\frac{b-a}{m+1}$, to the first term a. The *second mean* is obtained by augmenting the first by r, &c.

1. Find three arithmetical means between the extremes 2 and 18.

The formula $$r=\frac{b-a}{m+1}$$

gives $$r=\frac{18-2}{4}=4;$$

hence, the progression is

$$2 \,.\, 6 \,.\, 10 \,.\, 14 \,.\, 18.$$

2. Find twelve arithmetical means between 12 and 77.

The formula $$r=\frac{b-a}{m+1}$$

gives $$r=\frac{77-12}{13}=5.$$

Hence the progression is

$$12 \,.\, 17 \,.\, 22 \,.\, 27 \,.\,.\,.\,.\, 77$$

143. Remark. If the same number of arithmetical means are inserted between all the terms, taken two and two, these terms, and the arithmetical means united, will form but one and the same progression.

For, let $a \,.\, b \,.\, c \,.\, d \,.\, e \,.\, f \,.\,.\,.$ be the proposed progression, and m the number of means to be inserted between a and b, b and c, c and d . . .

From what has just been said, the common difference of each partial progression will be expressed by

$$\frac{b-a}{m+1}, \quad \frac{c-b}{m+1}, \quad \frac{d-c}{m+1} \ldots$$

which are equal to each other, since a, b, c . . . are in progression : therefore, the common difference is the same in each of the partial progressions ; and since the *last term* of the first, forms the *first term* of the second, &c, we may conclude that all of these partial progressions form a single progression.

EXAMPLES.

1. Find the sum of the first fifty terms of the progression 2 . 9 . 16 . 23 . . .

For the 50th term we have

$$l=2+49\times7=345.$$

Hence, $$S=(2+345)\times\frac{50}{2}=347\times25=8675.$$

2. Find the 100th term of the series 2 . 9 . 16 . 23 . . .
Ans. 695

3. Find the sum of 100 terms of the series 1 . 3 . 5 . 7 . 9 . . .
Ans. 10000

4. The greatest term is 70, the common difference 3, and the number of terms 21 : what is the least term and the sum of the series ?

Ans. Least term 10 ; sum of series 840

5 The first term is 4, the common difference 8, and the number of terms 8 : what is the last term, and the sum of the series ?

Ans. { Last term 60.
Sum = 256.

6. The first term is 2, the last term 20, and the number of terms 10 : what is the common difference ?

Ans. 2.

7. Insert four means between the two numbers 4 and 19 : what is the series ?

Ans. 4 . 7 . 10 : 13 . 16 . 19.

8. The first term of a decreasing arithmetical progression is 10, the common difference $\frac{1}{3}$, and the number of terms 21 : required the sum of the series.

Ans. 140.

9. In a progression by differences, having given the common difference 6, the last term 185, and the sum of the terms 2945 : find the first term, and the number of terms.

Ans. First term $=5$; number of terms 31.

10. Find nine arithmetical means between each antecedent and consequent of the progression 2 . 5 . 8 . 11 . 14 . . .

Ans. Common dif., or $r=0{,}3$.

11. Find the number of men contained in a triangular battalion, the first rank containing one man, the second 2, the third 3, and so on to the nth, which contains n. In other words, find the expression for the sum of the natural numbers 1, 2, 3 . . ., from 1 to n inclusively.

Ans. $S=\frac{n(n+1)}{2}$

12. Find the sum of the n first terms of the progression of uneven numbers 1, 3, 5, 7, 9 . . . *Ans.* $S=n^2$.

13. One hundred stones being placed on the ground in a straight line, at the distance of 2 yards from each other, how far will a person travel who shall bring them one by one to a basket, placed at 2 yards from the first stone?

Ans. 11 miles, 840 yards

Geometrical Proportion and Progression.

144. *Ratio* is the quotient arising from dividing one quantity by another quantity of the same kind. Thus, if the numbers 3 and 6 have the same unit, the ratio of 3 to 6 will be expressed by

$$\frac{6}{3}=2.$$

And in general, if A and B represent quantities of the same kind, the ratio of A to B will be expressed by

$$\frac{B}{A}.$$

145. If there be four numbers

2, 4, 8, 16,

having such values that the second divided by the first is equal to the fourth divided by the third, the numbers are

QUEST.—**144**. What is ratio? What is the ratio of 3 to 6? Of 4 to 12?

said to be in proportion. And in general, if there be four quantities, A, B, C, and D, having such values that

$$\frac{B}{A}=\frac{D}{C},$$

then A is said to have the same ratio to B that C has to D or, the ratio of A to B is equal to the ratio of C to D. When four quantities have this relation to each other, they are said to be in proportion. Hence, *proportion is an equality of ratios.*

To express that the ratio of A to B is equal to the ratio of C to D, we write the quantities thus:

$$A : B :: C : D;$$

and read, A is to B as C to D.

The quantities which are compared together are called the *terms* of the proportion. The first and last terms are called the *two extremes*, and the second and third terms, the *two means*. Thus, A and D are the extremes, and B and C the means.

146. Of four proportional quantities, the first and third are called the *antecedents*, and the second and fourth the *consequents;* and the last is said to be a fourth proportional to the other three taken in order. Thus, in the last proportion A and C are the antecedents, and B and D the consequents.

QUEST.—**145**. What is proportion? How do you express that four numbers are in proportion? What are the numbers called? What are the first and fourth called? What the second and third?—**146**. In four proportional quantities, what are the first and third called? What the second and fourth?

147. Three quantities are in proportion when the first has the same ratio to the second that the second has to the third; and then the middle term is said to be a mean proportional between the other two. For example,

$$3 : 6 :: 6 : 12;$$

and 6 is a mean proportional between 3 and 12.

148. Quantities are said to be in proportion by *inversion*, or *inversely*, when the consequents are made the antecedents and the antecedents the consequents.

Thus, if we have the proportion

$$3 : 6 :: 8 : 16,$$

the inverse proportion would be

$$6 : 3 :: 16 : 8.$$

149. Quantities are said to be in proportion by *alternation*, or *alternately*, when antecedent is compared with antecedent and consequent with consequent.

Thus, if we have the proportion

$$3 : 6 :: 8 : 16,$$

the alternate proportion would be

$$3 : 8 :: 6 : 16.$$

QUEST.—**147**. When are three quantities proportional? What is the middle one called?—**148**. When are quantities said to be in proportion by inversion, or inversely?—**149**. When are quantities in proportion by alternation?

150. Quantities are said to be in proportion by *composition*, when the sum of the antecedent and consequent is compared either with antecedent or consequent.

Thus, if we have the proportion

$$2 : 4 :: 8 : 16,$$

the proportion by composition would be

$$2+4 : 4 :: 8+16 : 16;$$

that is, $6 : 4 :: 24 : 16.$

151. Quantities are said to be in proportion by *division* when the difference of the antecedent and consequent is compared either with antecedent or consequent.

Thus, if we have the proportion

$$3 : 9 :: 12 : 36,$$

the proportion by division will be

$$9-3 : 9 :: 36-12 : 36;$$

that is, $6 : 9 :: 24 : 36.$

152. Equi-multiples of two or more quantities are the products which arise from multiplying the quantities by the same number.

Thus, if we have any two numbers, as 6 and 5, and multiply them both by any number, as 9, the equi-multiples will be 54 and 45; for

$$6\times9=54, \quad \text{and} \quad 5\times9=45.$$

QUEST.—**150**. When are quantities in proportion by composition?—**151**. When are quantities in proportion by division?—**152**. What are equi-multiples of two or more quantities?

Also, $m \times A$ and $m \times B$ are equi-multiples of A and B, the common multiplier being m.

153. Two quantities, A and B, are said to be *reciprocally proportional*, or *inversely proportional*, when one increases in the same ratio as the other diminishes. When this relation exists, either of them is equal to a constant quantity divided by the other.

Thus, if we had any two numbers, as 2 and 4, so related to each other that if we divided one by any number we must multiply the other by the same number, one would increase just as fast as the other would diminish, and their product would be constant.

154. If we have the proportion

$$A : B :: C : D,$$

we have $$\frac{B}{A}=\frac{D}{C}, \text{ (Art. } \mathbf{145}\text{)};$$

and by clearing the equation of fractions, we have

$$BC=AD.$$

That is, *Of four proportional quantities, the product of the two extremes is equal to the product of the two means.*

This general principle is apparent in the proportion between the numbers

$$2 : 10 :: 12 : 60,$$

which gives $$2 \times 60=10 \times 12=120.$$

QUEST.—**153**. When are two quantities said to be reciprocally proportional?—**154**. If four quantities are proportional, what is the product of the two means equal to?

155. If four quantities, A, B, C, D, are so related to each other that

$$A \times D = B \times C,$$

we shall also have $$\frac{B}{A} = \frac{D}{C};$$

and hence, $$A : B :: C : D.$$

That is: *If the product of two quantities is equal to the product of two other quantities, two of them may be made the extremes, and the other two the means of a proportion.*

Thus, if we have

$$2 \times 8 = 4 \times 4,$$

we also have

$$2 : 4 :: 4 : 8.$$

156. If we have three proportional quantities

$$A : B :: B : C,$$

we have $$\frac{B}{A} = \frac{C}{B};$$

hence, $$B^2 = AC.$$

That is: *The square of the middle term is equal to the product of the two extremes.*

Thus, if we have the proportion

$$3 : 6 :: 6 : 12,$$

we shall also have

$$6 \times 6 = 6^2 = 3 \times 12 = 36.$$

QUEST.—**155.** If the product of two quantities is equal to the product of two other quantities, may the four be placed in a proportion? How? —**156.** If three quantities are proportional, what is the product of the extremes equal to?

157. If we have

$$A : B :: C : D, \text{ and consequently } \frac{B}{A}=\frac{D}{C},$$

multiply both members of the last equation by $\frac{C}{B}$, we then obtain,

$$\frac{C}{A}=\frac{D}{B},$$

and hence, $A : C :: B : D.$

That is: *If four quantities are proportional, they will be in proportion by alternation.*

Let us take, as an example,

10 : 15 : : 20 : 30.

We shall have, by alternating the terms,

10 : 20 : : 15 : 30.

158. If we have

$$A : B :: C : D \text{ and } A : B :: E : F,$$

we shall also have

$$\frac{B}{A}=\frac{D}{C} \text{ and } \frac{B}{A}=\frac{F}{E};$$

hence, $\frac{D}{C}=\frac{F}{E}$ and $C : D :: E : F.$

That is: *If there are two sets of proportions having an*

QUEST.—**157**. If four quantities are proportional, will they be in proportion by alternation?

antecedent and consequent in the one equal to an antecedent and consequent of the other, the remaining terms will be proportional.

If we have the two proportions

$$2 : 6 :: 8 : 24 \quad \text{and} \quad 2 : 6 :: 10 : 30,$$

we shall also have

$$8 : 24 :: 10 : 30.$$

159. If we have

$$A : B :: C : D, \quad \text{and consequently} \quad \frac{B}{A}=\frac{D}{C},$$

we have, by dividing 1 by each member of the equation,

$$\frac{A}{B}=\frac{C}{D}, \quad \text{and consequently} \quad B : A :: D : C.$$

That is: *Four proportional quantities will be in proportion, when taken inversely.*

To give an example in numbers, take the proportion

$$7 : 14 :: 8 : 16;$$

then, the inverse proportion will be

$$14 : 7 :: 16 : 8,$$

in which the ratio is one-half.

160. The proportion

$$A : B :: C : D \quad \text{gives} \quad A \times D = B \times C.$$

QUEST.—**158.** If you have two sets of proportions having an antecedent and consequent in each, equal; what will follow?—**159.** If four quantities are in proportion, will they be in proportion when taken inversely?

To each member of the last equation add $B \times D$. We shall then have

$$(A+B) \times D = (C+D) \times B;$$

and by separating the factors, we obtain

$$A+B : B :: C+D : D.$$

If, instead of adding, we subtract $B \times D$ from both members, we have

$$(A-B) \times D = (C-D) \times B;$$

which gives

$$A-B : B :: C-D : D.$$

That is: *If four quantities are proportional, they will be in proportion by composition or division.*

Thus, if we have the proportion

$$9 : 27 :: 16 : 48,$$

we shall have, by composition,

$$9+27 : 27 :: 16+48 : 48:$$

that is, $36 : 27 :: 64 : 48,$

in which the ratio is three-fourths.

The proportion gives us, by division,

$$27-9 : 27 :: 48-16 : 48;$$

that is, $18 : 27 :: 32 : 48,$

in which the ratio is one and one-half.

QUEST.—**160**. If four quantities are in proportion, will they be in proportion by composition? Will they be in proportion by division? What is the difference between composition and division?

161. If we have

$$\frac{B}{A}=\frac{D}{C},$$

and multiply the numerator and denominator of the first member by any number m, we obtain

$$\frac{mB}{mA}=\frac{D}{C} \quad \text{and} \quad mA : mB :: C : D.$$

That is : *Equal multiples of two quantities have the same ratio as the quantities themselves.*

For example, if we have the proportion

$$5 : 10 :: 12 : 24,$$

and multiply the first antecedent and consequent by 6, we have

$$30 : 60 :: 12 : 24,$$

in which the ratio is still 2.

162. The proportions

$$A : B :: C : D \quad \text{and} \quad A : B :: E : F,$$

give $\quad A \times D = B \times C \quad$ and $\quad A \times F = B \times E$;

adding and subtracting these equations, we obtain

$$A(D \pm F) = B(C \pm E), \quad \text{or} \quad A : B :: C \pm E : D \pm F$$

That is : *If* C *and* D, *the antecedent and consequent, be augmented or diminished by quantities* E *and* F, *which have the same ratio as* C *to* D, *the resulting quantities will also have the same ratio.*

QUEST.—**161.** Have equal multiples of two quantities the same ratio as the quantities?—**162.** Suppose the antecedent and consequent be augmented or diminished by quantities having the same ratio?

Let us take, as an example, the proportion

$$9 : 18 :: 20 : 40,$$

in which the ratio is 2.

If we augment the antecedent and consequent by 15 and 30 which have the same ratio, we shall have

$$9+15 : 18+30 :: 20 : 40;$$

that is, $24 : 48 :: 20 : 40,$

in which the ratio is still 2.

If we diminish the second antecedent and consequent by the same numbers, we have

$$9 : 18 :: 20-15 : 40-30;$$

that is, $9 : 18 :: 5 : 10,$

in which the ratio is still 2.

163. If we have several proportions

$A : B :: C : D,$ which gives $A\times D=B\times C,$

$A : B :: E : F,$ „ „ $A\times F=B\times E,$

$A : B :: G : H,$ „ „ $A\times H=B\times G.$

&c, &c,

we shall have, by addition,

$$A(D+F+H)=B(C+E+G);$$

and by separating the factors,

$$A : B :: C+E+G : D+F+H.$$

That is: *In any number of proportions having the same ratio, any antecedent will be to its consequent, as the sum of the antecedents to the sum of the consequents.*

Let us take, for example,

$$2 : 4 :: 6 : 12 \quad \text{and} \quad 1 : 2 :: 3 : 6, \quad \&c$$

Then, $\quad 2 : 4 :: 6+3 : 12+6;$

that is, $\quad 2 : 4 :: 9 : 18,$

in which the ratio is still 2.

164. If we have four proportional quantities

$$A : B :: C : D, \quad \text{we have} \quad \frac{B}{A}=\frac{D}{C};$$

and raising both members to any power, as n, we have

$$\frac{B^n}{A^n}=\frac{D^n}{C^n},$$

and consequently

$$A^n : B^n :: C^n : D^n.$$

That is : *If four quantities are proportional, any like powers or roots will be proportional.*

If we have, for example,

$$2 : 4 :: 3 : 6,$$

we shall have $\quad 2^2 : 4^2 :: 3^2 : 6^2;$

that is, $\quad 4 : 16 :: 9 : 36,$

in which the terms are proportional, the ratio being 4.

165. Let there be two sets of proportions,

$$A : B :: C : D, \quad \text{which gives} \quad \frac{B}{A}=\frac{D}{C},$$

$$E : F :: G : H, \quad \text{"} \quad \text{"} \quad \frac{F}{E}=\frac{H}{G}.$$

QUEST.—**163.** In any number of proportions having the same ratio, how will any one antecedent be to its consequent ?—**164.** In four proportional quantities, how are like powers or roots ?

Multiply them together, member by member, we have

$$\frac{BF}{AE}=\frac{DH}{CG} \quad \text{which gives} \quad AE : BF :: CG : DH.$$

That is : *In two sets of proportional quantities, the products of the corresponding terms will be proportional.*

Thus, if we have the two proportions

$$8 : 16 :: 10 : 20$$

and

$$3 : 4 :: 6 : 8,$$

we shall have

$$24 : 64 :: 60 : 160.$$

Geometrical Progression.

166. We have thus far only required that the ratio of the first term to the second should be the same as that of the third to the fourth.

If we impose the farther condition, that the ratio of the second term to the third shall also be the same as that of the first to the second, or of the third to the fourth, we shall have a series of numbers, each one of which, divided by the preceding one, will give the same ratio. Hence, if any term be multiplied by this quotient, the product will be the succeeding term. A series of numbers so formed is called a *geometrical progression*. Hence,

A *Geometrical Progression*, or *progression by quotients*, is a series of terms, each of which is equal to the product of

QUEST.—**165.** In two sets of proportions, how are the products of the corresponding terms ?

that which precedes it by a *constant number*, which number is called the *ratio* of the progression. Thus,

$$1 : 3 : 9 : 27 : 81 : 243, \&c,$$

is a geometrical progression, which is written by merely placing two dots between each two of the terms. Also,

$$64 : 32 : 16 : 8 : 4 : 2 : 1$$

is a geometrical progression, in which the ratio is *one-half*.

In the first progression each term is contained three times in the one that follows, and hence the ratio is 3. In the second, each term is contained one-half times in the one which follows, and hence the ratio is one-half.

The first is called an *increasing* progression, and the second a *decreasing* progression.

Let a, b, c, d, e, f, . . . be numbers in a progression by quotients; they are written thus:

$$a : b : c : d : e : f : g \ldots$$

and it is enunciated in the same manner as a progression by differences. It is necessary, however, to make the distinction, that one is a series of equal differences, and the other a series of equal quotients or ratios. It should be remarked that each term is at the same time an antecedent and a consequent, except the first, which is only an antecedent, and the last, which is only a consequent.

QUEST.—**166.** What is a geometrical progression? What is the ratio of the progression? If any term of a progression be multiplied by the ratio, what will the product be? If any term be divided by the ratio, what will the quotient be? How is a progression by quotients written? Which of the terms is only an antecedent? Which only a consequent? How may each of the others be considered?

167. Let q denote the ratio of the progression

$$a : b : c : d \ldots;$$

q being >1 when the progression is *increasing*, and $q<1$ when it is *decreasing*. Then, since

$$\frac{b}{a}=q, \quad \frac{c}{b}=q, \quad \frac{d}{c}=q, \quad \frac{e}{d}=q, \text{ \&c,}$$

we have

$$b=aq, \quad c=bq=aq^2, \quad d=cq=aq^3, \quad e=dq=aq^4,$$
$$f=eq=aq^5 \ \ldots;$$

that is, the second term is equal to aq, the third to aq^2, the fourth to aq^3, the fifth to aq^4, &c; and in general, any term n, that is, one which has $n-1$ terms before it, is expressed by aq^{n-1}.

Let l be this term; we then have the formula

$$l=aq^{n-1},$$

by means of which we can obtain any term without being obliged to find all the terms which precede it. Hence, to find the last term of a progression, we have the following

RULE.

I. *Raise the ratio to a power whose exponent is one less than the number of terms.*

II. *Multiply the power thus found by the first term: the product will be the required term.*

QUEST.—167. By what letter do we denote the ratio of the progression? In an increasing progression is q greater or less than 1? In a decreasing progression is q greater or less than 1? If a is the first term and q the ratio, what is the second term equal to? What the third? What the fourth? What is the last term equal to? Give the rule for finding the last term

EXAMPLES.

1. Find the 5th term of the progression

$$2 : 4 : 8 : 16 \;. \;.$$

in which the first term is 2 and the common ratio 2.

$$5\text{th term} = 2 \times 2^4 = 2 \times 16 = 32 \quad \textit{Ans.}$$

2. Find the 8th term of the progression

$$2 : 6 : 18 : 54 \;. \;. \;.$$

$$8\text{th term} = 2 \times 3^7 = 2 \times 2187 = 4374 \quad \textit{Ans.}$$

3. Find the 6th term of the progression

$$2 : 8 : 32 : 128 \;. \;. \;.$$

$$6\text{th term} = 2 \times 4^5 = 2 \times 1024 = 2048 \quad \textit{Ans.}$$

4. Find the 7th term of the progression

$$3 : 9 : 27 : 81 \;. \;. \;.$$

$$7\text{th term} = 3 \times 3^6 = 3 \times 729 = 2187 \quad \textit{Ans.}$$

5. Find the 6th term of the progression

$$4 : 12 : 36 : 108 \;. \;. \;.$$

$$6\text{th term} = 4 \times 3^5 = 4 \times 243 = 972 \quad \textit{Ans.}$$

6. A person agreed to pay his servant 1 cent for the first day, two for the second, and four for the third, doubling every day for ten days: how much did he receive on the tenth day? *Ans.* $5,12.

7. What is the 8th term of the progression

$$9 : 36 : 144 : 576 \dots$$

$$\text{8th term}=9\times4^7=9\times16384=147456 \quad \textit{Ans.}$$

8 Find the 12th term of the progression

$$64 : 16 : 4 : 1 : \frac{1}{4} \dots$$

$$\text{12th term}=64\left(\frac{1}{4}\right)^{11}=\frac{4^3}{4^{11}}=\frac{1}{4^8}=\frac{1}{65536} \quad \textit{Ans.}$$

168. We will now proceed to determine the sum of n terms of the progression

$$a : b : c : d : e : f : \dots : i : k : l;$$

l denoting the nth term.

We have the equations (Art. **167**),

$$b=aq, \quad c=bq, \quad d=cq, \quad e=dq, \dots k=iq, \quad l=kq;$$

and by adding them all together, member to member, we deduce

Sum of 1st members. *Sum of 2nd members*

$$b+c+d+e+\dots+k+l=(a+b+c+d+\dots+i+k)q;$$

in which we see that the first member wants the first term a, and the polynomial within the parenthesis in the second member wants the last term l. Hence, if we call the sum of the terms S, we have

$$S-a=(S-l)q=Sq-lq, \quad \text{or} \quad Sq-S=lq-a;$$

whence $$S=\frac{lq-a}{q-1}.$$

22

Therefore, to obtain the sum of the terms of a geometrical progression, we have the following

RULE.

I. *Multiply the last term by the ratio.*

II. *Subtract the first term from the product.*

III. *Divide the remainder by the ratio diminished by unity, and the quotient will be the sum of the series.*

1. Find the sum of eight terms of the progression

$$2 : 6 : 18 : 54 : 162 \ldots 2\times 3^7=4374.$$

$$S=\frac{lq-a}{q-1}=\frac{13122-2}{2}=6560.$$

2. Find the sum of the progression

$$2 : 4 : 8 : 16 : 32.$$

$$S=\frac{lq-a}{q-1}=\frac{64-2}{1}=62$$

3. Find the sum of ten terms of the progression

$$2 : 6 : 18 : 54 : 162 \ldots 2\times 3^9=39366.$$

Ans. 59048.

4. What debt may be discharged in a year, or twelve months, by paying \$1 the first month, \$2 the second month,

---

QUEST.—**168.** Give the rule for finding the sum of the series. What is the first step? What the second? What the third?

$4 the third month, and so on, each succeeding payment being double the last; and what will be the last payment?

Ans. { Debt, . . $4095.
Last payment, $2048.

5. A gentleman married his daughter on New Year's day, and gave her husband 1*s*. towards her portion, and was to double it on the first day of every month during the year: what was her portion? *Ans.* £204 15*s*.

6. A man bought 10 bushels of wheat on the condition that he should pay 1 cent for the 1st bushel, 3 for the second, 9 for the third, and so on to the last: what did he pay for the last bushel and for the ten bushels?

Ans. { Last bushel, $196,83.
Total cost, $295,24.

7. A man plants 4 bushels of barley, which, at the first harvest, produced 32 bushels; these he also plants, which, in like manner, produce 8 fold; he again plants all his crop, and again gets 8 fold, and so on for 16 years: what is his last crop, and what the sum of the series?

Ans. { Last, 140737488355328*bu*.
Sum, 160842843834660.

169 When the progression is decreasing, we have $q<1$ and $l<a$; the above formula

$$S=\frac{lq-a}{q-1},$$

for the sum is then written under the form

$$S=\frac{a-lq}{1-q},$$

in order that the two terms of the fraction may be positive.

QUEST.—**163**. What is the formula for the sum of the series of a decreasing progression?

1. Find the sum of the terms of the progression

$$32 : 16 : 8 : 4 : 2.$$

$$S=\frac{a-lq}{1-q}=\frac{32-2\times\frac{1}{2}}{\frac{1}{2}}=\frac{31}{\frac{1}{2}}=62.$$

2. Find the sum of the first twelve terms of the progression

$$64 : 16 : 4 : 1 : \frac{1}{4} : \ldots : 64\left(\frac{1}{4}\right)^{11}, \text{ or } \frac{1}{65536}.$$

$$S=\frac{a-lq}{1-q}=\frac{64-\frac{1}{65536}\times\frac{1}{4}}{\frac{3}{4}}=\frac{256-\frac{1}{65536}}{3}=85+\frac{65535}{196608}.$$

REMARK.—**170.** We perceive that the principal difficulty consists in obtaining the numerical value of the last term, a tedious operation, even when the number of terms is not very great.

3. Find the sum of 6 terms of the progression

$$512 : 128 : 32 \ldots$$

Ans. $682\frac{1}{2}$

4. Find the sum of seven terms of the progression

$$2187 : 729 : 243 \ldots$$

Ans. 3279.

5. Find the sum of six terms of the progression

$$972 : 324 : 108$$

Ans. 1456.

6. Find the sum of 8 terms of the progression

$$147456 : 36864 : 9216 \ldots$$

Ans. 196605.

Of Progressions having an infinite number of terms

171. Let there be the decreasing progression

$$a : b : c : d : e : f : \ldots$$

containing an indefinite number of terms. In the formula

$$S=\frac{a-lq}{1-q},$$

substitute for l its value aq^{n-1} (Art. **167**), and we have

$$S=\frac{a-aq^n}{1-q},$$

which represents the sum of n terms of the progression. This may be put under the form

$$S=\frac{a}{1-q}-\frac{aq^n}{1-q}.$$

Now, since the progression is decreasing, q is a proper fraction; and q^n is also a fraction, which diminishes as n increases. Therefore, the greater the number of terms we take, the more will $\frac{a}{1-q}\times q^n$ diminish, and consequently the more will the partial sum of these terms approximate to an equality with the first part of S, that is, to $\frac{a}{1-q}$. Finally, when n is taken greater than any given number, or n=infinity, then $\frac{a}{1-q}\times q^n$ will be less than any given number, or will become equal to 0; and the expression $\frac{a}{1-q}$ will represent the true value of the sum of all the terms of the series. Whence we may conclude, that the expression for *the sum of the terms of a decreasing progression, in which the number of terms is infinite, is*

$$S=\frac{a}{1-q}.$$

That is, *equal to the first term divided by* 1 *minus the ratio.*

This is, properly speaking, the *limit* to which the *partial sums* approach, by taking a greater number of terms in the progression. The difference between these sums and $\frac{a}{1-q}$ can become as small as we please, and will only become *nothing* when the number of terms taken is infinite

EXAMPLES.

1. Find the sum of

$$1 : \frac{1}{3} : \frac{1}{9} : \frac{1}{27} : \frac{1}{81} \text{ to infinity.}$$

We have for the expression of the sum of the terms

$$S=\frac{a}{1-q}=\frac{1}{1-\frac{1}{3}}=\frac{3}{2}. \quad \textit{Ans.}$$

The error committed by taking this expression for the value of the sum of the n first terms, is expressed by

$$\frac{a}{1-q}\times q^n=\frac{3}{2}\left(\frac{1}{3}\right)^n.$$

First take $n=5$; it becomes

$$\frac{3}{2}\left(\frac{1}{3}\right)^5=\frac{1}{2\ .\ 3^4}=\frac{1}{162}$$

QUEST.—**165**. When the progression is decreasing and the number of terms infinite, what is the value of the sum of the series?

When $n=6$, we find

$$\frac{3}{2}\left(\frac{1}{3}\right)^6=\frac{1}{162}\times\frac{1}{3}=\frac{1}{486}.$$

Whence we see that the *error committed*, when $\frac{3}{2}$ is taken for the sum of a certain number of terms, is less in proportion as this number is greater.

2. Again take the progression

$$1 : \frac{1}{2} : \frac{1}{4} : \frac{1}{8} : \frac{1}{16} : \frac{1}{32} : \&c. \ldots$$

We have $$S=\frac{a}{1-q}=\frac{1}{1-\frac{1}{2}}=2. \quad Ans.$$

3. What is the sum of the progression

$$1, \frac{1}{10}, \frac{1}{100}, \frac{1}{1000}, \frac{1}{10000}, \text{ \&c, to infinity.}$$

$$S=\frac{1}{1-q}=\frac{1}{1-\frac{1}{10}}=1\frac{1}{9} \quad Ans.$$

172. In the several questions of geometrical progression there are five numbers to be considered:

1st. The first term, ... a.
2nd. The ratio, ... q.
3rd. The number of terms, ... n.
4th. The last term, ... l.
5th. The sum of the terms, ... S.

QUEST.—**166**. How many numbers are considered in geometrical progression? What are they?

173. We shall terminate this subject by the question,

To find a mean proportional between any two numbers, as m and n.

Denote the required mean by x. We shall then have (Art. **156**),

$$x^2 = m \times n,$$

and hence $$x = \sqrt{m \times n}.$$

That is, *Multiply the two numbers together, and extract the square root of the product.*

1. What is the geometrical mean between the numbers 2 and 8?

$$\text{Mean} = \sqrt{8 \times 2} = \sqrt{16} = 4 \quad \textit{Ans.}$$

2. What is the mean between 4 and 16? *Ans.* 8

3. What is the mean between 3 and 27? *Ans.* 9.

4. What is the mean between 2 and 72? *Ans.* 12.

5. What is the mean between 4 and 64? *Ans.* 16.

QUEST.—**167**. How do you find a mean proportional between two numbers?

CHAPTER VIII.

Of Logarithms.

174. The nature and properties of the logarithms in common use, will be readily understood, by considering attentively the different powers of the number 10. They are,

$10^0 = 1$

$10^1 = 10$

$10^2 = 100$

$10^3 = 1000$

$10^4 = 10000$

$10^5 = 100000$

&c. &c.

It is plain that the *indices* or *exponents* 0, 1, 2, 3, 4, 5, &c. form an arithmetical series of which the common difference is 1; and that the numbers 1, 10, 100, 1000, 10000, 100000, &c. form a geometrical series of which the common ratio is 10. The number 10, is called the *base* of the system of logarithms; and the *indices* 0, 1, 2, 3, 4, 5, &c., are the loga-

QUEST.—**174.** What relation exists between the exponents 1, 2, 3, &c.? How are the corresponding numbers 10, 100, 1000? What is the common difference of the exponents? What is the common ratio of the corresponding numbers? What is the base of the common system of logarithms? What are the indices? Of what number is the index 1 the logarithm? The index 2? The index 3?

rithms of the numbers which are produced by raising 10 to the powers denoted by those indices.

175. If we denote the base of the system by a, and the logarithm of any number by m, then the number itself will e the mth power of a: that is, if we represent the corres-onding number by M,

$$a^m = M$$

Thus, if we make $m=0$, M will be equal to 1; if $m=1$, M will be equal to 10, &c. Hence,

The logarithm of a number is the exponent of the power to which it is necessary to raise the base of the system in order to produce the number.

176. Letting, as before, a denote the base of the system of logarithms, m any exponent, and M the corresponding number: we shall then have,

$$a^m = M$$

in which m is the logarithm of M.

If we take a second exponent n, and let N denote the corresponding number, we shall have,

$$a^n = N$$

in which n is the logarithm of N.

If now, we multiply the first of these equations by the second, member by member, we have

$$a^m \times a^n = a^{m+n} = M \times N;$$

but since a is the base of the system, $m+n$ is the logarithm $M \times N$; hence,

---

QUEST.—**175.** If we denote the base of a system by a, and the exponent by m, what will represent the corresponding number? What is the logarithm of a number? **176.** To what is the sum of the logarithms of any two numbers equal? To what then, will the addition of logarithms correspond?

The sum of the logarithms of any two numbers is equal to the logarithm of their product.

Therefore, *the addition of logarithms corresponds to the multiplication of their numbers.*

177. If we divide the equations by each other, member by member, we have,

$$\frac{a^m}{a^n}=a^{m-n}=\frac{M}{N};$$

but since a is the base of the system, $m-n$ is the logarithm of $\frac{M}{N}$; hence,

If one number be divided by another, the logarithm of the quotient will be equal to the logarithm of the dividend diminished by that of the divisor.

Therefore, *the subtraction of logarithms corresponds to the division of their numbers.*

178. Let us examine further the equations

$$10^0=1$$
$$10^1=10$$
$$10^2=100$$
$$10^3=1000$$
&c. &c.

It is plain that the logarithm of 1 is 0, and that the logarithms of all the numbers between 1 and 10, are greater than 0 and less than 1. They are generally expressed by decimal fractions: thus,

$$\log 2=0.301030$$

QUEST.—**177** If one number be divided by another, what will the logarithm of the quotient be equal to? To what then will the subtraction of logarithms correspond? **178.** What is the logarithm of 1? Between what limits are the logarithms of all numbers between 1 and 10? How are they generally expressed?

The logarithms of all the numbers greater than 10 and less than 100, are greater than 1 and less than 2, and are generally expressed by 1 and a decimal fraction: thus,

$$\log 50 = 1.698970$$

The part of the logarithm which stands on the left of the decimal point, is called the *characteristic* of the logarithm. The characteristic is always *one less than the places of integer figures in the number whose logarithm is taken.*

Thus, in the first case, for numbers between 1 and 10, there is but one place of figures, and the characteristic is 0. For numbers between 10 and 100, there are two places of figures, and the characteristic is 1; and similarly for other numbers.

Table of Logarithms.

179. A table of logarithms is a table in which are written the logarithms of all numbers between 1 and some other given number. A table showing the logarithms of the numbers between 1 and 100 is annexed. The numbers are written in the column designated by the letter N, and the logarithms in the columns designated by Log.

QUEST.—How is it with the logarithms of numbers between 10 and 100? What is that part of the logarithm called which stands at the left of the characteristic? What is the value of the characteristic? **179.** What is a table of logarithms? Explain the manner of finding the logarithms of numbers between 1 and 100?

TABLE.

N.	Log.	N.	Log.	N.	Log.	N.	Log.
1	0.000000	26	1.414973	51	1.707570	76	1.880814
2	0.301030	27	1.431364	52	1.716003	77	1.886491
3	0.477121	28	1.447158	53	1.724276	78	1.892095
4	0.602060	29	1.462398	54	1.732394	79	1.897627
5	0.698970	30	1.477121	55	1.740363	80	1.903090
6	0.778151	31	1.491362	56	1.748188	81	1.908485
7	0.845098	32	1.505150	57	1.755875	82	1.913814
8	0.903090	33	1.518514	58	1.763428	83	1.919078
9	0.954243	34	1.531479	59	1.770852	84	1.924279
10	1.000000	35	1.544068	60	1.778151	85	1.929419
11	1.041393	36	1.556303	61	1.785330	86	1.934498
12	1.079181	37	1.568202	62	1.792392	87	1.939519
13	1.113943	38	1.579784	63	1.799341	88	1.944483
14	1.146128	39	1.591065	64	1.806180	89	1.949390
15	1.176091	40	1.602060	65	1.812913	90	1.954243
16	1.204120	41	1.612784	66	1.819544	91	1.959041
17	1.230449	42	1.623249	67	1.826075	92	1.963788
18	1.255273	43	1.633468	68	1.832509	93	1.968483
19	1.278754	44	1.643453	69	1.838849	94	1.973128
20	1.301030	45	1.653213	70	1.845098	95	1.977724
21	1.322219	46	1.662758	71	1.851258	96	1.982271
22	1.342423	47	1.672098	72	1.857333	97	1.986772
23	1.361728	48	1.681241	73	1.863323	98	1.991226
24	1.380211	49	1.690196	74	1.869232	99	1.995635
25	1.397940	50	1.698970	75	1.875061	100	2.000000

EXAMPLES.

1. Let it be required to multiply 8 by 9, by means of loga rithms. We have seen, Art. 176, that the sum of the logarithms is equal to the logarithm of the product. Therefore, find the logarithm of 8 from the table, which is 0.903090 and then the logarithm of 9, which is 0.954243; and their sum, which is 1.857333, will be the logarithm of the product.

In searching along in the table, we find that 72 stands opposite this logarithm: hence, 72 is the product of 8 by 9.

2. What is the product of 7 by 12?

Logarithm of 7 is, ...	0.845098
Logarithm of 12 is, ...	1.079181
Logarithm of their product, . .	1.924279

and the number corresponding is 84.

3. What is the product of 9 by 11?

Logarithm of 9 is, . . .	0.954243
Logarithm of 11 is, ...	1.041393
Logarithm of their product, . .	1.995636

and the corresponding number is 99.

4. Let it be required to divide 84 by 3. We have seen in Article 177, that the subtraction of Logarithms corresponds to the division of their numbers. Hence, if we find the logarithm of 84, and then subtract from it the logarithm of 3, the remainder will be the logarithm of the quotient.

The logarithm of 84 is, . . .	1.924279
The logarithm of 3 is, . . .	0.477121
Their difference is, ...	1.447158

and the number corresponding is 28.

5. What is the product of 6 by 7?

Logarithm of 6 is, ...	0.778151
Logarithm of 7 is, ...	0.845098
Their sum is, ...	1.623249

and the corresponding number of the table, 42.

SUPPLEMENT.

EXAMPLES IN ADDITION AND SUBTRACTION

1. What is the sum of

$$ax^n+bx^n+cx^n+dx^n$$

2. What is the sum of

$$ax^n+bx^n-cx^n-dx^n$$

3. What is the sum of

$$10a^4+3a^4+6a^4-a^4-5a^4$$

4. What is the sum of

$$5a^3-7a^3+11a^3+a^4$$

5. What is the sum of

$$a^nb^m-9a^m+5a^nb^m+6a^m+10a^nb^m$$

6. What is the sum of

$$5a^3b^2+7ab^2c-3a^mb^5-12ab^2c+6a^3b^2-9a^3b^3+b^x-8a^mb^5-3b^x$$

7. What is the sum of

$$5a^4b+3a^2b^2c-7ab-6a^4b+2a^2b^2c+17ab+9a^4b-8a^2b^2c$$
$$-10ab$$

8. What is the sum of

$$5a^mb^p+3a^3b^{m-1}-3a^3--3ca^mb^p+4g^2a^3b^{m-1}-a+10a^3+a^mb^p+$$
$$a+3a^2b^2-2g^2a^3b^{m-1}$$

9. What is the sum of

$$9a^3b^2c^4-7b+18b-5a^nb^m+c^x-3d^5+3a^nb^m-ha^3b^2c^4+3c^x$$
$$-5d^5$$

10. From $\quad -9a^mx^2-13+2ab^3x-4b^mcx^2$

take $\quad 3b^mcx^2-9a^mx^2-6+2ab^3x$

11. From $5a^4-7a^3b^2-3cd^2+7d$
take $-15a^3b^2+3a^4-3a^2-7cd^2$

12. From $9a^mb^2+6ab^m-d^5+18a^4b^n$
take $7a^4b^n+d^5-8ab^m+9a^mb^2$

13. From $12b^5d^6-16a^8b^6-5a^mb^n+6ac^b$
take $6a^mb^n-6ac^b+16a^8b^6+12b^5d^6$

14. From $8a^3b^3c^5-12a^mb+6ax^4+8ad^m$
take $8ad^m-8a^3b^3c^5-12a^mb+6ax^4$

15. From $12a^mb^n-9ax^5-4ab+6a^2b^2-a$
take $3a-6a^2b^2+12a^mb^n-9ax^5+5ab$

EXAMPLES IN MULTIPLICATION.

1. What is the product of
$$a^m \times a^n$$

2. What is the product of
$$2a^3 \times 7a^9 \times -3a^6$$

3. What is the product of
$$a^5b \times a^7d \times 10a \times 6a^2 \times -1$$

4. What is the product of
$$-a^{p-q} \times -3a^{p-4} \times f \times 5a^{3q+7}cx$$

5. What is the product of
$$5a^3b^4 \times 10a^2b^5c \times -3a^7$$

6. What is the product of
$$-7a^5b^6c^2 \times 8a^5b^2d \times 7b^8c \times -1$$

7. What is the product of
$$a^mb^pcq \times a^nb^rc^q \times a^mb \times -a$$

8. What is the product of
$$(a^2-3ab-5b^2) \times 4a^2b$$

9. What is the product of
$$(2a^3b^5-5a^2c^6+9a^3b^2c^3) \times 3a^2bc^2$$

10. What is the product of

$$(7h^5l+2l^3-3ah^3l^2+7)\times-8h^4l^5$$

11. What is the product of

$$(a^3b^4-cb^5d^3f+3c^m)\times-2bc^2d$$

12. What is the product of

$$(a^{m-1}b^{2m+3}-6a^{-m-3}b^p+ab^m)\times a^{3m+2}b^{m-1}.$$

13. What is the product of

$$(3k^2-5kl+2l^2)\times(k^2-7kl)$$

14. What is the product of

$$(6f^2-17fl+3l^2)\times(f^5+4f^4l)$$

15. What is the product of

$$(4a^2-16ax+3x^2)\times(5a^3-2a^2x).$$

16. What is the product of

$$(a^2+a^4+a^6)\times(a^2-1)$$

17. What is the product of

$$(a^4-2a^3b+4a^2b^2-8ab^3+16b^4)\times(a+2b).$$

18. What is the product of

$$(2a^4x^2-3b^4y^2)\times(2a^4x^2+3b^4y^2)$$

19. What is the product of

$$(7a^3-5a^2b+6ab^2-2b^3)\times(3a^4-4a^3b+16a^2b^2)$$

20. What is the product of

$$(a^6-3a^4b^2+5a^2b^4)\times(7a^4-4a^2b^2+b^4).$$

EXAMPLES IN DIVISION.

1. Divide a^m by a^n.

2. Divide a^m by a^{2n}.

3. Divide $8a^{16}$ by $2a^4$.

4. Divide ca^{18} by da^4.

5. Divide $6(a+b)^9$ by $3(a+b)^8$.

6. Divide $(a+x)^2\times(a+y)^3$ by $(a+x)\times(a+y)^2$.

7. Divide $6a^3b^2-15a^2f+27a^4bx$, by $3a^2$.

8. Divide bc^3-c^3x by $b-x$.

9. Divide $a^3+a^2b-ab^2-b^3$ by $a-b$.

10. Divide $3a^5+16a^4b-33a^3b^2+14a^2b^3$ by a^2+7ab.

11. Divide $a^7-6a^6b^3+14a^5b^6-12a^4b^9$ by $a^3-2a^2b^3$.

12. Divide $a^4-2a^2b^2+b^4$ by a^2-b^2.

13. Divide $-a^8b^4+15a^{11}b^5-48a^{14}b^5-20a^{17}b^7$
by $10a^9b^2-a^6b$.

14. Divide a^8-16z^8 by a^4-2z^4.

15. Divide $2a^4-13a^3b+31a^2b^2-38ab^3+24b^4$
by $2a^2-3ab+4b^2$.

16. Divide $4c^4-9b^2c^2+6b^3c-b^4$ by $2c^2-3bc+b^2$.

17. Divide $-1+a^3n^3$ by $-1+an$.

18. Divide $a^6+2a^3z^3+z^6$ by a^2-az+z^2.

19. Divide $\frac{1}{3}-6z^2+27z^4$ by $\frac{1}{3}+2z+3z^2$.

20. Divide $a^6-16a^3x^3+64x^6$ by $a^2-4ax+4x^2$.

21. Divide $a^3d^3-3a^2cd^3+3ac^2a^3-c^3d^3+a^2c^2d^2-ac^3d^2$
by $a^2d^2-2acd^2+c^2d^2+ac^2d$.

EXAMPLES IN REDUCTION OF FRACTIONS.

1. Reduce to its simplest terms the fraction

$$\frac{18acf-6bdcf-2ad}{3adf}$$

2. Reduce to its simplest terms the fraction

$$\frac{8a^2-6ab+4c+1}{-2a}$$

3. Reduce to its simplest terms the fraction

$$\frac{12acfg-4af^2g+3fg^2h}{4a^2b^2fg}.$$

4. Reduce to its simplest terms the fraction

$$\frac{ab-ac}{b-c}$$

5. Reduce $a-b+\frac{cx}{x-a}$ to the form of a fraction.

6. Reduce $x-\frac{b-y}{ax-c}$ to the form of a fraction.

7. Reduce $a+b+\frac{y-a}{d}$ to the form of a fraction.

8. Reduce $x-ab-\frac{c-d}{a-x}$ to the form of a fraction.

9. Reduce $a-\frac{b+c-d}{x-y}$ to the form of a fraction.

10. Reduce $6af^2x+9af-\frac{b+ax}{b}$ to the form of a fraction.

11. Reduce $5acx-y-\frac{x-ax}{fac}$ to the form of a fraction.

12. Reduce to an entire or mixed quantity the fraction

$$\frac{6a^3b^2-10a^2f+7a^4bx}{2a^2}.$$

13. Reduce to an entire or mixed quantity the fraction

$$\frac{\frac{3}{4}a^2x^5-\frac{2}{9}ax^3+3abx}{\frac{2}{3}a^2x^3}.$$

14. Reduce to an entire or mixed quantity the fraction

$$\frac{a^3-2a^2x+ab+ax^2-bx}{a-x}$$

15. Reduce the following fractions to a common denominator: viz.

$$\frac{a-x}{a+x},\ \frac{b-c}{f},\ \frac{4ax-c}{a-x}.$$

16. Reduce the following fractions to a common denominator: viz.

$$\frac{a}{3ax-b},\ \frac{a-x}{3b}\ \text{and}\ -\frac{c}{a-x}$$

17. Reduce the following fractions to a common denominator: viz.

$$\frac{4af-x}{7a-c},\ -\frac{a-x}{c}\ \text{and}\ \frac{5ac}{y}$$

18. Reduce the following fractions to a common denominator: viz.

$$\frac{4ax+b}{8ac-f}\ \text{and}\ \frac{8ac-f}{4ax}$$

19. Reduce the following fractions to a common denominator: viz.

$$-\frac{a+x}{a-x},\ \frac{a-b}{a+x}\ \text{and}\ \frac{c-d}{a}$$

20. Reduce the following fractions to a common denominator: viz.

$$\frac{a+b-c}{x-a},\ \frac{c+f}{c}\ \text{and}\ \frac{x-a}{x+a}.$$

ADDITION, SUBTRACTION, MULTIPLICATION AND DIVISION OF FRACTIONS.

1. What is the sum of $\frac{a}{x-a}$, $\frac{b}{a-x}$ and c

2. What is the sum of $\frac{a-x}{b}$, $\frac{c-a}{a-c}$ and y

3. What is the sum of $\frac{3a}{b}$, $\frac{x-ay}{d}$, $\frac{4ax}{3c-f}$,

4. What is the sum of $\frac{ax-f}{c+a}$, $\frac{ac-g}{-x}$, $5ay$

5. What is the sum of $3a+\frac{2ax}{x-a}$, $2a-\frac{-x}{a+b}$.

6. From $8a+\frac{3a}{b}$ take $\frac{6a-x}{a-x}$

7. From $\frac{8x-ax}{b}$ take $3-\frac{5ax}{c}$

8. From $\frac{ax+3ay}{acx-ay}$ take $\frac{3ax-af}{ay}$

9. From $ay-\frac{ax+d}{ax}$ take $3ay+\frac{6ax-y}{9}$

10. From $\frac{5ay}{8x}$ take $7ab^2-\frac{6ax-ay}{8a-x}$

11. Multiply $7a+\frac{3b}{6}$ by $\frac{a-x}{x+a}$

12. Multiply $\frac{5a}{x-a}$ by $3ay-\frac{5ax-x^2}{x}$

13 Multiply $\frac{9a-x}{3a+x}$ by $2a+\frac{6ax-x^2}{9a}$

14. Multiply $3ax+\frac{5ay-x}{-y}$ by $5+\frac{a}{b}$

15. Multiply $6a+\frac{5-8a}{-b}$ by $\frac{6a-b}{a^2-b^2}$

16. Divide $3ac-2adc-f+\frac{c}{d}$ by $2a$

17. Divide $\frac{a}{d}+\frac{fd}{2c}-3ac+7$ by $\frac{3c}{d}$

18. Divide $3a^2-\frac{7ab}{2}-\frac{21ac}{4}-\frac{5b^2}{2}+\frac{83bc}{8}-\frac{3c^2}{2}$ by $3a-5b+\frac{3c}{4}$

19. Divide $2f^2-\frac{55fh}{12}+\frac{29fx}{9}+\frac{21h^2}{8}-\frac{15hx}{4}+\frac{x^2}{3}$ by $\frac{2f}{3}-\frac{3h}{4}+x$

20. Divide $-\frac{5x^2}{9}+\frac{11xy}{3}-\frac{10xz}{3}+\frac{15y^2}{4}+25yz$ by $-\frac{2x}{3}+5y$.

EXAMPLES IN EQUATIONS OF THE FIRST DEGREE.

1. Given $\left\{\begin{array}{l} x-\frac{1}{2}y+\frac{4x}{5}=6\frac{4}{5} \\ \frac{x-y}{2}+7x=41 \end{array}\right\}$ to find x and y.

2 Given $\left\{\begin{array}{l}\frac{x-y}{4}+\frac{x+y}{5}=2\frac{1}{10} \\ \frac{1}{2}x-y+4\frac{1}{4}y=12\frac{1}{4}\end{array}\right\}$ to find x and y.

3 Given $ax+\frac{x}{c-a}+\frac{3}{4}(x-f)=g$ to find x.

4. Given $\frac{a-x}{4}+\frac{b-x}{5}-\frac{a-x}{c}=d$, to find x.

5. Given $\left\{\begin{array}{l}\frac{3y-x}{6}+\frac{2x-y}{4}=5 \\ 6x-y+\frac{8-2x}{4}=43\frac{1}{2}\end{array}\right\}$ to find x and y.

6. Given $\left\{\begin{array}{l}\frac{3x-8}{4}+\frac{y-6}{5}+y=18\frac{7}{10} \\ 8x-3-\frac{6-y}{3}=79\end{array}\right\}$ to find x and y.

7. Given $\left\{\begin{array}{l}\frac{4x-4}{3}-\frac{y-5}{4}+6=12\frac{2}{3} \\ \frac{1}{2}x-\frac{1}{3}y+\frac{y-4}{3}=\frac{5}{3}.\end{array}\right\}$ to find x and y.

8. Given $\frac{a-x}{3}-\frac{a-2x}{3}+x=b$, to find x.

9. Given $\frac{3a-6x}{b}-\frac{2a-3x}{c}+x-d=f$, to find x.

10. Given $\left\{\begin{array}{l}ax-by=c \\ a-y+x=d\end{array}\right\}$ to find x and y.

11 Given $\frac{x-5}{3}+\frac{x-1}{2}+4(x-3)=68$, to find x.

12. Given $\left\{\begin{array}{l} \frac{1}{2}x + \frac{1}{3}y + \frac{1}{4}z = 6 \\ 8-y + \frac{6-z}{3} + 1 = 3\frac{2}{3} \\ x-y + \frac{z-x}{2} = 0. \end{array}\right\}$ to find x, y, and z.

13. Given $\left\{\begin{array}{l} x + 2y + 3z = 14 \\ x-y + z = 2 \\ 3x + 6y + z = 18 \end{array}\right\}$ to find x, y, and z.

14. Given $\left\{\begin{array}{l} \frac{1}{3}x - \frac{1}{8}y + \frac{1}{6}z = 2\frac{1}{3} \\ \frac{x-y}{4} + \frac{x+z}{5} = 4\frac{3}{10} \\ \frac{x-3}{2} + \frac{y-z}{4} = 3. \end{array}\right\}$ to find x, y and z.

15 Given $\left\{\begin{array}{l} 13x+7y-341=7\frac{1}{2}y+43\frac{1}{2}x \\ 2x+\frac{1}{2}y=1 \end{array}\right\}$ to find x and y.

16. Given $\left\{\begin{array}{l} (x+5)(y+7)=(x+1)(y-9)+112 \\ 2x+10 = 3y+1 \end{array}\right\}$ to find x and y.

17. Given $\left\{\begin{array}{l} ax = by \\ x+y = c \end{array}\right\}$ to find x and y.

18. Given $\left\{\begin{array}{l} ax + by = c \\ fx + gy = h \end{array}\right\}$ to find x and y.

19. Given $\left\{\begin{array}{l} \frac{a}{b+y} = \frac{b}{3a+x} \\ ax+2by = d \end{array}\right\}$ to find x and y.

20. Given $\left\{\begin{array}{l} bcx = cy-2b \\ b^2y + \frac{a(c^3-b^3)}{bc} = \frac{2b^3}{c} + c^2x \end{array}\right\}$ to find x and y.

21. Given $\left\{\begin{array}{l} 3x+5y = \frac{(8b-2f)bf}{b^2-f^2} \\ y-x = -\frac{2bf^2}{b^2-f^2} \end{array}\right\}$ to find x and y.

22. Given $\left\{\begin{array}{l} x + y + z = 29\frac{1}{4} \\ x + y - z = 18\frac{1}{4} \\ x - y + z = 13\frac{3}{4} \end{array}\right\}$ to find x, y and z.

23. Given $\left\{\begin{array}{l} 3x + 5y = 161 \\ 7x + 2z = 209 \\ 2y + z = 89 \end{array}\right\}$ to find x, y and z.

24. Given $\left\{\begin{array}{l} \frac{1}{x} + \frac{1}{y} = a \\ \frac{1}{x} + \frac{1}{z} = b \\ \frac{1}{y} + \frac{1}{z} = c \end{array}\right\}$ to find x, y and z.

25. Given $\left\{\begin{array}{l} ax + by = c \\ dx + ey = f \\ gy + hz = l \end{array}\right\}$ to find x, y and z.

EXAMPLES IN EQUATIONS OF THE SECOND DEGREE.

1. Given $x^2 - 5\frac{3}{4}x = 18$ to find x.

2. Given $3x^2 - 2x = 65$ to find x.

3. Given $622x = 15x^2 + 6384$ to find x.

4. Given $11\frac{3}{4}x - 3\frac{1}{2}x^2 = -41\frac{1}{4}$, to find x.

5. Given $9\frac{1}{3}x^2 - 90\frac{1}{3}x + 195 = 0$, to find x.

6. Given $20748 - 1616x + 21x^2 = 0$, to find x.

7. Given $9\frac{1}{3}x^2 - 90\frac{1}{3}x + 195 = 0$, to find x.

8. Given $\frac{18x^2}{5} + \frac{18078x}{65} + 4728 = 0$, to find x.

9. Given $x^2-8x=14$ to find x.

10. Given $3x^2+x=7$ to find x.

11. Given $118x-2\frac{1}{8}x^2=20$ to find x.

12. Given $6x-30=3x^2$ to find x.

13. Given $8x^2-7x+34=0$

14. Given $4x^2-9x=5x^2-255\frac{3}{4}-8x$ to find x.

15. Given $80x+\frac{3x^2}{4}+\frac{21x-27782}{12}=1859\frac{1}{3}-3x^2$

16. Given $\frac{31}{6x}=\frac{16}{117-2x}+1$ to find x.

17. Given $\frac{25x+180}{10x-81}=\frac{40x}{5x-8}-\frac{3}{5}$ to find x.

18. Given $\frac{18+x}{6(3-x)}=\frac{20x+9}{19-7x}-\frac{65}{4(3-x)}$ to find x.

19. Given $x^2-7x+3\frac{1}{4}=0$

20. Given $4x^2-9x=5x^2-255\frac{3}{4}-8x$ to find x.

21. Given $\frac{x}{x+60}=\frac{7}{3x-5}$ to find x.

22. Given $\frac{40}{x-5}+\frac{27}{x}=13$ to find x.

23. Given $\frac{8x}{x+2}-6=\frac{20}{3x}$ to find x.

24. Given $\frac{48}{x+3}=\frac{165}{x+10}-5$ to find x.

PROMISCUOUS QUESTIONS.

EQUATIONS OF THE FIRST DEGREE.

1. A person expends 30 cents for apples and pears, giving one cent for four apples, and one cent for five pears: he then sold, at the prices he gave, half his apples and one-third his pears, for 13 cents. How many did he buy of each?

2. A tailor cut 19 yards from each of three equal pieces of cloth, and 17 yards from another of the same length, and found that the four remnants were altogether equal to 142 yards. How many yards in each piece?

3. A fortress is garrisoned by 2600 men, consisting of infantry, artillery, and cavalry. Now, there are nine times as many infantry, and three times as many artillery soldiers, as there are cavalry. How many are there of each corps?

4. All the journeyings of an individual amounted to 2970 miles. Of these he travelled $3\frac{1}{2}$ times more by water than on horseback, and $2\frac{1}{4}$ times more on foot than by water. How many miles did he travel in each way?

5. A sum of money was divided between two persons, A and B. A's share was to exceed B's in the proportion of 5 to 3, and to exceed five-ninths of the entire sum by 50. What was the share of each?

6. There are 52 pieces of money in each of two bags, out of which A and B help themselves. A takes twice as much

as B left, and B takes seven times as much as A left. How much did each take?

7. Two persons, A and B, agree to purchase a house to gether, worth $1200. Says A to B, give me two-thirds of your money and I can purchase it alone; but, says B to A, if you give me three-fourths of your money I shall be able to purchase it alone. How much had each?

8. A father directs that $1170 shall be divided among his three sons, in proportion to their ages. The oldest is twice as old as the youngest, and the second is one-third older than the youngest. How much was each to receive?

9. Three regiments are to furnish 594 men, and each to furnish in proportion to its strength. Now, the strength of the first is to the second as 3 to 5; and that of the second to the third as 8 to 7? How many must each furnish?

10. A grocer finds that if he mixes sherry and brandy in the proportion of 2 to 1, the mixture will be worth 78*s.* per dozen; but if he mixes them in the proportion of 7 to 2, he can get 79*s.* a dozen. What is the price of each liquor per dozen?

11. A person bought 7 books, the prices of which were in arithmetical progression, (in shillings.) The price of the one next above the cheapest, was 8 shillings, and the price of the dearest, 23 shillings. What was the price of each book?

12. A number consists of three digits, which are in arithmetical proportion. If the number be divided by the sum of the digits, the quotient will be 26; but, if 198 be added to it, the digits will be inverted.

13. A person has three horses, and a saddle which is worth $220. If the saddle be put on the back of the first horse, it

will make his value equal to that of the second and third; if it be put on the back of the second, it will make his value double that of the first and third; if it be put on the back of the third, it will make his value triple that of the first and second. What is the value of each horse?

14. The crew of a ship consisted of her complement of sailors, and a number of soldiers. There are 22 sailors to every three guns, and 10 over; also, the whole number of hands is five times the number of soldiers and guns together. But after an engagement, in which the slain were one-fourth of the survivors, there wanted 5 men to make 13 men to every two guns. Required, the number of guns, soldiers, and sailors.

15. Three persons have $96, which they wish to divide equally between them. In order to do this, A, who has the most, gives to B and C as much as they have already: then B divides with A and C in the same manner, that is, by giving to each as much as he had after A had divided with them: C then makes a division with A and B, when it is found that they all have equal sums. How much had each at first?

16. To divide the number a into three such parts, that the first shall be to the second as m to n, and the second to the third as p to q.

17. Five heirs, A, B, C, D, and E, are to divide an inheritance of $5600. B is to receive twice as much as A, and $200 more; C three times as much as A, less $400; D the half of what B and C receive together, and 150 more; and E the fourth part of what the four others get, plus $475. How much did each receive?

18. A person has four casks, the second of which being

filled from the first, leaves the first four-sevenths full. The third being filled from the second, leaves it one-fourth full, and when the third is emptied into the fourth, it is found to fill only nine-sixteenths of it. But the first will fill the third and fourth, and leave 15 quarts remaining. How many quarts does each hold?

19. A courier who had started from a place 10 days, was pursued by a second courier. The first travels 4 miles a day, the other 9. How many days before the second will overtake the first?

20. If the first courier had left n days before the other, and made a miles a day, and the second courier had travelled b miles, how many days before the second would have overtaken the first?

21. A courier goes $31\frac{1}{2}$ miles every five hours, and is followed by another after he had been gone eight hours. The second travels $22\frac{1}{2}$ miles every three hours. How many hours before he will overtake the first?

22. Two places are eighty miles apart, and a person leaves one of them and travels towards the other, at the rate of $3\frac{1}{2}$ miles per hour. Eight hours after, a person departs from the second place, and travels at the rate of $5\frac{1}{8}$ miles per hour. How long before they will meet each other?

23. Three masons, A, B, and C, are to build a wall. A and B together can do it in 12 days; B and C in 20 days; and A and C in 15 days. In what time can each do it alone, and in what time can they all do it if they work together?

24. A laborer can do a certain work expressed by a, in a time expressed by b; a second laborer, the work c in a time d; a third, the work e in a time f. It is required to find the

time it would take the three laborers, working together, to perform the work g.

25. Required to find three numbers with the following conditions. If 6 be added to the 1st and 2d, the sums are to one another as 2 to 3. If 5 be added to the 1st and 3d, the sums are as 7 to 11; but, if 36 be subtracted from the 2d and 3d, the remainders will be as 6 to 7.

26. The sum of $500 was put out at interest, in two separate sums, the smaller sum at two per cent. more than the other. The interest of the larger sum was afterwards increased, and that of the smaller diminished, by one per cent. By this, the interest of the whole was augmented one-fourth. But if the interest of the greater sum had been so increased, without any diminution of the less, the interest of the whole would have been increased one-third. What were the sums, and what the rate per cent.?

27. The ingredients of a loaf of bread weighing 15*lbs.* are rice, flour, and water The weight of the rice, augmented by 5*lbs.*, is two-thirds the weight of the flour; and the weight of the water is one-fifth the weight of the flour and rice together. Required, the weight of each.

28. Several detachments of artillery divided a certain number of cannon balls. The first took 72 and $\frac{1}{9}$ of the remainder; the next 144 and $\frac{1}{9}$ of the remainder; the third 216 and $\frac{1}{9}$ of the remainder; the fourth 288 and $\frac{1}{9}$ of what was left; and so on, until nothing remained; when it was found that the balls were equally divided. Required, the number of balls and the number of detachments

29. A banker has two kinds of money; it takes a pieces

of the first to make a crown, and b of the second to make the same sum. He is offered a crown for c pieces. How many of each kind must he give?

30. Find what each of three persons, A, B, and C is worth, knowing, 1st, that what A is worth, added to l times what B and C are worth, is equal to p; 2d, that what B is worth, added to m times what A and C are worth, is equal to q; 3d, that what C is worth, added to n times what A and B are worth, is equal to r.

31. Find the values of the estates of six persons, A, B, C, D, E, and F, from the following conditions. 1st. The sum of the estates of A and B is equal to a; that of C and D to b; and that of E and F to c. 2d. The estate of A is worth m times that of C; the estate of D is worth n times that of E, and the estate of F is worth p times that of B.

PROMISCUOUS QUESTIONS.

INVOLVING EQUATIONS OF THE SECOND DEGREE.

1. FIND three numbers, such, that the difference between the third and second shall exceed the difference between the second and first by 6: that the sum of the numbers shall be 33, and the sum of their squares 467.

2. It is required to find three numbers in geometrical progression, such that their sum shall be 14, and the sum of their squares 84.

3. What two numbers are those, whose sum multiplied by the greater, gives 144, and whose difference multiplied by the less, gives 14?

4. What two numbers are those, which are to each other as m to n, and the sum of whose squares is b?

5. What two numbers are those, which are to each other as m to n, and the difference of whose squares is b?

6. A certain capital is out at 4 per cent. interest. If we multiply the number of dollars in the capital by the number of dollars in the interest, for five months, we obtain \$117041$\frac{2}{3}$. What is the capital?

7. A person has three kinds of goods, which together cost \$230$\frac{5}{24}$. One pound of each article costs as many times $\frac{1}{24}$ of a dollar as there are pounds of that article. Now, he has

one-third more of the second kind than of the first, and $3\frac{1}{3}$ times more of the third than of the second. How many pounds had he of each?

8. Required to find three numbers, such, that the product of the first and second shall be equal to a; the product of the first and third equal to b; and the sum of the squares of the second and third equal to c.

9. It is required to find three numbers, whose sum shall be 38, the sum of their squares 634, and the difference between the second and first greater by 7 than the difference between the third and second.

10. Find three numbers in geometrical progression, whose sum shall be 52, and the sum of the extremes to the mean, as 10 to 3.

11. The sum of three numbers in geometrical progression is 13, and the product of the mean by the sum of the extremes is 30. What are the numbers?

12. It is required to find three numbers, such, that the product of the first and second, added to the sum of their squares, shall be 37; and the product of the first and third, added to the sum of their squares, shall be 49; and the product of the second and third, added to the sum of their squares, shall be 61.

14. Find two numbers, such, that their difference, added to the difference of their squares, shall be equal to 150, and whose sum, added to the sum of their squares, shall be equal to 330.

15. It is required to find a number consisting of three digits, such, that the sum of the squares of the digits shall be

104; the square of the middle digit to exceed twice the product of the other two by 4; and if 594 be subtracted from the number, the three digits become inverted.

16. The sum of two numbers and the sum of their squares being given, to find the numbers.

17. The sum, and the sum of the cubes, of two numbers being given, to find the numbers.

18. To find three numbers in arithmetical progression such, that their sum shall be equal to 18, and the product of the two extremes added to 25 shall be equal to the square of the mean.

19. Having given the sum, and the sum of the fourth powers of two numbers; to find the numbers.

20. To find three numbers in arithmetical progression, such, that the sum of their squares shall be equal to 1232, and the square of the mean greater than the product of the two extremes, by 16.

21. To find two numbers whose sum is 80, and such, that if they be divided alternately by each other, the sum of the quotients shall be $3\frac{1}{3}$.

22. To find two numbers whose difference shall be 10, and if 600 be divided by each of them, the difference of the quotients shall also be 10.

www.ingramcontent.com/pod-product-compliance
Lightning Source LLC
LaVergne TN
LVHW010218110826
845151LV00004B/1134

* 9 7 8 1 4 2 5 5 2 6 9 4 8 *